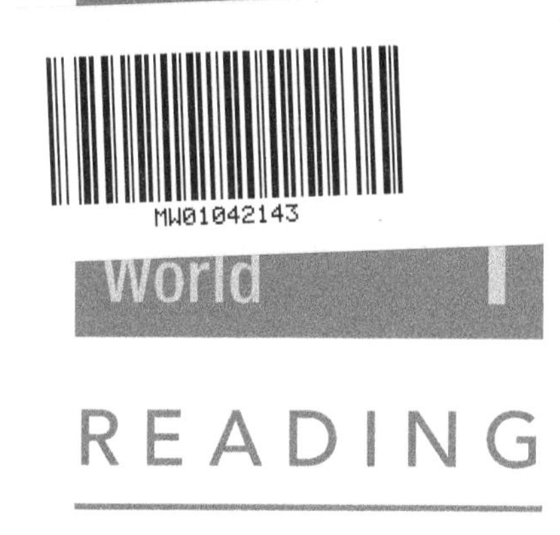

Academic Encounters

2nd Edition

Jennifer Wharton
Series Editor: Bernard Seal

CAMBRIDGE
UNIVERSITY PRESS

32 Avenue of the Americas, New York, NY 10013-2473, USA

Cambridge University Press is part of the University of Cambridge.

It furthers the University's mission by disseminating knowledge in the pursuit of education, learning and research at the highest international levels of excellence.

www.cambridge.org
Information on this title: www.cambridge.org/9781107683631

First published 2009
Second edition 2013
5th printing 2016

Printed in Italy by Rotolito Lombarda S.p.A.

A catalog record for this publication is available from the British Library

Library of Congress Cataloging in Publication Data

Wharton, Jennifer, 1968-.
Academic encounters : the natural world : reading, writing / Jennifer Wharton. -- Second edition.
pages cm. -- (Academic Encounters)
ISBN 978-1-107-68363-1 (student's ed. : level 2) -- ISBN 978-1-107-69450-7 (Teacher's ed. : level 2)
1. English language--Rhetoric--Problems, exercises etc. 2. Study skills--Problems, exercises, etc.
3. Earth sciences--Problems, exercises, etc. 4. Readers (Secondary) I. Title.

PE1128.W57 2013
428.2'4--dc23

2013005887

ISBN 978-1-107-68363-1 Student's Book
ISBN 978-1-107-69450-7 Teacher's Manual

Additional resources for this publication at www.cambridge.org/academicencounters

Art direction and layout services: Kamae Design, Oxford, UK
Photo research: Suzanne Williams

Table of Contents

Scope and sequence iv

Introduction viii

To the student xv

Acknowledgments xvi

Unit 1: Planet Earth 1

Chapter 1 The Physical Earth 4

Chapter 2 The Dynamic Earth 27

Unit 2: Water on Earth 51

Chapter 3 Earth's Water Supply 54

Chapter 4 Earth's Oceans 77

Unit 3: The Air Around Us 101

Chapter 5 Earth's Atmosphere 104

Chapter 6 Weather and Climate 127

Unit 4: Life on Earth 151

Chapter 7 Plants and Animals 154

Chapter 8 Humans 179

Weights and Measures 205

Appendix: Academic Word List vocabulary 206

Skills index 207

Credits 208

Scope and sequence

Unit 1: Planet Earth • 1

	Content	Reading Skills	Writing Skills
Chapter 1 The Physical Earth page 3	**Reading 1** Our Solar System **Reading 2** Earth's Four Systems **Reading 3** Rocks on Our Planet	Thinking about the topic Previewing art Asking and answering questions about a text Previewing key parts of a text	Parts of speech Comparative adjectives
Chapter 2 The Dynamic Earth page 27	**Reading 1** Plate Tectonics **Reading 2** Volcanoes **Reading 3** Earthquakes	Using headings to remember main ideas Building background knowledge about the topic Reading boxed texts Illustrating main ideas Thinking about the topic Reading for main ideas	Writing simple and compound sentences Writing definitions Pronoun reference Showing contrast

Unit 2: Water on Earth • 51

	Content	Reading Skills	Writing Skills
Chapter 3 Earth's Water Supply page 54	**Reading 1** The Water Cycle **Reading 2** Groundwater and Surface Water **Reading 3** Glaciers	Thinking about the topic Examining graphics Sequencing Reading about statistics Increasing reading speed Reading for main ideas Scanning	Identifying topic sentences Identifying topic sentences and supporting sentences Writing topic sentences and supporting sentences
Chapter 4 Earth's Oceans page 77	**Reading 1** Oceans **Reading 2** Currents **Reading 3** Waves and Tsunamis	Thinking about the topic Building background knowledge about the topic Reading maps Examining graphics Brainstorming Reading for main ideas and details	Writing about superlatives Describing results Concluding sentences Parallel structure Both…and and neither…nor Reviewing paragraph structure

Vocabulary Skills	Academic Success Skills	Learning Outcomes
Words from Latin and Greek Cues for finding word meaning Learning verbs with their prepositions	Highlighting Making a pie chart Answering multiple-choice questions Labeling diagrams	Write an academic paragraph about a place on Earth you like
Previewing key words Prefixes Prepositional phrases Using grammar, context, and background knowledge to guess meaning	Reading maps Answering true/false questions	

Vocabulary Skills	Academic Success Skills	Learning Outcomes
Antonyms Suffixes that change verbs into nouns Countable and uncountable nouns Subject-verb agreement	Understanding test questions Answering multiple-choice questions Mapping Conducting a survey	Write an academic paragraph about a water feature on earth
Subject-verb agreement Too and very Adjective suffixes	Taking notes Highlighting Labeling a map Organizing ideas	

Unit 3: The Air Around Us • 101

	Content	Reading Skills	Writing Skills
Chapter 5 Earth's Atmosphere page 104	**Reading 1** The Composition of the Atmosphere **Reading 2** The Structure of the Atmosphere **Reading 3** Clouds	Previewing key terms Building background knowledge about the topic Thinking about the topic Previewing key parts of a text Examining graphics Previewing art	Reviewing paragraph structure Transition words Writing about height Writing an observation report
Chapter 6 Weather and Climate page 127	**Reading 1** Climates Around the World **Reading 2** Storms **Reading 3** Hurricanes	Thinking about the topic Applying what you have read Previewing key parts of a text Increasing reading speed Reading for main ideas	Introducing examples

Unit 4: Life on Earth • 151

	Content	Reading Skills	Writing Skills
Chapter 7 Plants and Animals page 154	**Reading 1** Living Things **Reading 2** Plant Life **Reading 3** Animal Life	Thinking about the topic Building background knowledge about the topic Previewing key parts of a text	Writing about similarities Writing about differences Writing about similarities and differences
Chapter 8 Humans page 179	**Reading 1** The Brain **Reading 2** The Skeletal and Muscular Systems **Reading 3** The Heart and the Circulatory System	Thinking about the topic Applying what you have read Increasing reading speed Asking and answering questions about a text Scanning for details Building background knowledge about the topic Sequencing	Writing a description Writing about the body

Vocabulary Skills	Academic Success Skills	Learning Outcomes
Guessing meaning from context Describing parts Playing with words Colons, such as, and lists Words from Latin and Greek When clauses	Examining test questions Taking notes with a chart Using symbols and abbreviations	Write an academic paragraph about the climate in a place you know
Defining key words Using a dictionary Using this/that/these/those to connect ideas Synonyms Prepositions of location	Understanding averages Using a Venn diagram to organize ideas from a text Examining statistics Thinking critically about the topic	

Vocabulary Skills	Academic Success Skills	Learning Outcomes
Word families Defining key words Cues for finding word meaning That clauses Compound words	Answering true/false questions Asking for clarification Conducting a survey Making an outline Applying what you have read Thinking critically about the topic	Write an academic paragraph about the human body
Using adjectives Gerunds Words that can be used as nouns or verbs Prepositions of direction Playing with words	Highlighting and taking notes Using a dictionary Conducting an experiment Answering multiple-choice questions Highlighting and making an outline	

Academic Encounters: Preparing Students for Academic Coursework

The Series

Academic Encounters is a sustained content-based series for English language learners preparing to study college-level subject matter in English. The goal of the series is to expose students to the types of texts and tasks that they will encounter in their academic coursework and provide them with the skills to be successful when that encounter occurs.

Academic Content

At each level in the series, there are two thematically paired books. One is an academic reading and writing skills book, in which students encounter readings that are based on authentic academic texts. In this book, students are given the skills to understand texts and respond to them in writing. The reading and writing book is paired with an academic listening and speaking skills book, in which students encounter interview and lecture material specially prepared by experts in their field. In this book, students learn how to take notes from a lecture, participate in discussions, and prepare short oral presentations.

Flexibility

The books at each level may be used as stand-alone reading and writing books or listening and speaking books. They may also be used together to create a complete four-skills course. This is made possible because the content of each book at each level is very closely related. Each unit and chapter, for example, has the same title and deals with similar content, so that teachers can easily focus on different skills, but the similar content, as they toggle from one book to the other. Additionally, if the books are taught together, when students are presented with the culminating unit writing or speaking assignment, they will have a rich and varied supply of reading and lecture material to draw on.

A Sustained Content Approach

A sustained content approach teaches language through the study of subject matter from one or two related academic content areas. This approach simulates the experience of university courses and better prepares students for academic study.

Students benefit from a sustained content approach

Real-world academic language and skills

Students learn how to understand and use academic language because they are studying actual academic content.

An authentic, intensive experience

By immersing students in the language of a single academic discipline, sustained content helps prepare them for the rigor of later coursework.

Natural recycling of language

Because a sustained content course focuses on a particular academic discipline, concepts and language naturally recur. As students progress through the course, their ability to work with authentic language improves dramatically.

Knowledge of common academic content

When students work with content from the most popular university courses, they gain real knowledge of these academic disciplines. This helps them to be more successful when they move on to later coursework.

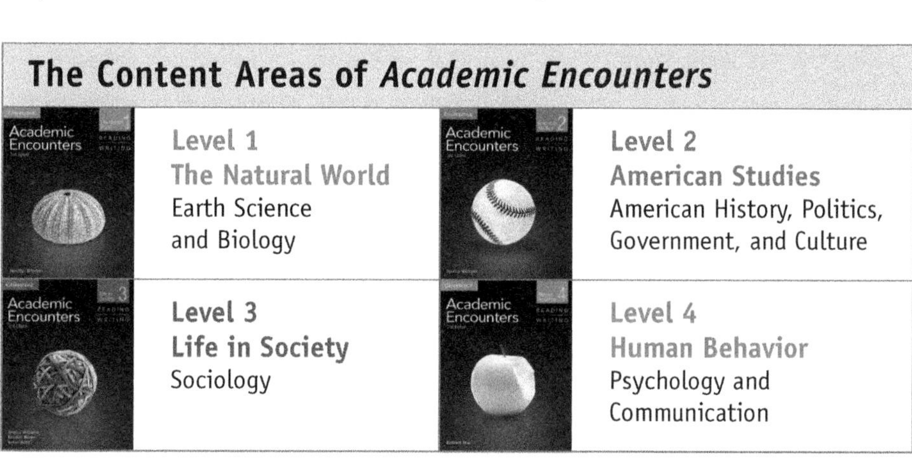

The Content Areas of *Academic Encounters*

Level 1	**The Natural World** Earth Science and Biology
Level 2	**American Studies** American History, Politics, Government, and Culture
Level 3	**Life in Society** Sociology
Level 4	**Human Behavior** Psychology and Communication

Academic Skills

Academic Encounters teaches skills in four main areas. A set of icons highlights which skills are practiced in each exercise.

R Reading Skills

The reading skills tasks are designed to help students develop strategies before reading, while reading, and after reading.

Writing Skills

Students learn how to notice and analyze written texts, develop critical writing skills, and apply these in longer writing tasks. These skills and tasks were carefully selected to prepare students for university study.

Vocabulary Development

Vocabulary learning is an essential part of improving one's ability to read an academic text. Tasks throughout the books focus on particular sets of vocabulary that are important for reading in a specific subject area as well as vocabulary from the Academic Word List.

Academic Success

Besides learning how to read, write, and build their language proficiency, students also have to learn other skills that are particularly important in academic settings. These include skills such as learning how to prepare for a content test, answering certain types of test questions, taking notes, and working in study groups.

Learning to read academic content

PREPARING TO READ

1 Thinking about the topic

Look at these photographs. Then discuss the questions below in a small group.

a.

b.

c.

1. Do you know the names or locations of any of the places in the photographs?
2. What do they all have in common?
3. What are some other famous places made of rock?
4. What do people use rocks for? Try to think of at least three uses.
5. Did builders use any rocks to construct your school or your home? Are there any things inside your school or home that are made of rock?

2 Previewing key parts of a text

A Read these key parts of the text "Rocks on Our Planet" on pages 19–20:

- the title
- the introductory paragraphs
- the headings
- the photographs and illustrations

B Answer these questions with a partner.

1. How many main types of rocks does Earth have?
2. What are their names?
3. What is the rock cycle?

18 Unit 1 *Planet Earth*

Pre-reading is a crucial step in the reading process. *Academic Encounters* teaches important skills to **help students succeed** when preparing to read university textbooks.

Reading 3

EARTHQUAKES

As 1974 came to an end and the new year began, animals in Haicheng, China, started acting strangely. Snakes normally hibernate underground during the winter, but they suddenly came out of their holes. Dogs began to bark and run around wildly, and horses became so upset that some ran away. Why were the animals acting like this? Many people think that the animals sensed what was coming: On February 4, 1975, the earth began to shake and buildings collapsed as a very large earthquake struck the city of Haicheng.

What causes earthquakes?

When the tectonic plates that make up Earth's crust move past each other, they often bump or rub against each other. The earth, or ground, above the plates moves as well. This movement is called an earthquake. Sometimes the plates get stuck. The pressure increases as the two plates try to move past each other but cannot. They finally move with a sudden and powerful jerk. This can also cause an earthquake. During a small earthquake, the ground simply shakes a little, and people may not even notice. However, a strong movement can cause the ground to shake and roll violently. It can make buildings and bridges fall. It can also cause the ground to split open and form a large **fault**, or crack.

fault a large break in the surface of the earth

Where do earthquakes happen?

Earthquakes can happen anywhere, but certain places have more earthquakes. These places sit on tectonic plates that move frequently. One example is the area around the Pacific Plate, which includes China, the Philippines, Japan, and the western coasts of Canada, the United States, and South America. Earthquakes are common in those places. The deadliest earthquake in modern times happened in 1976 in Tangshan, China. It lasted less than two minutes, but more than 250,000 people died, and more than 90 percent of the buildings collapsed. Earthquake scientists study places such as Tangshan because of the many faults in these areas and the activity of the tectonic plates.

San Andreas fault in California

40 Unit 1 *Planet Earth*

The readings are based on **university textbooks**, so students improve their ability to **read authentic academic materials.**

The Structure of Academic Text

3 Subject-verb agreement V

Singular subjects take singular verbs. Remember that, in the present tense, third-person singular verbs end in *-s* or *-es*.

subject / verb
Big Rock **weighs** almost 14,970 metric tons. **It is** 9 meters high.

subject / verb
A glacier **carves** deep valleys as **it slides** over the ground.

Plural subjects take plural verbs. Plural verbs do not end in *-s*.

subject / verb
Scientists **study** glaciers to learn about world climates.

Uncountable nouns take singular verbs.

subject / verb
Water **covers** 97 percent of Earth's surface.

Complete the sentences with the correct form of the verb in parentheses.

1. A glacier ________ (form) when the snow in an area does not melt. Over time, this snow ________ (turn) into ice.
2. In some glaciers, the ice ________ (be) thousands of years old.
3. Glaciers ________ (contain) more than 75 percent of Earth's freshwater.
4. Every continent ________ (have) glaciers except Australia.

4 Identifying topic sentences and supporting sentences W R

Many paragraphs in academic English begin with a **topic sentence**. **Supporting sentences** follow the topic sentence.

- **topic sentences:** state the main idea of a text
- **major supporting sentences:** directly support a topic sentence; answer questions such as *how*, *why*, or *when*
- **minor supporting sentences:** directly support a major supporting sentence; answer questions such as *how*, *why*, or *when* with specific information

A Read the paragraph below.

Glaciers change the surface of our planet in different ways. One way is by shaping the land. For example, glaciers carve U-shaped valleys and form sharp mountaintops. Glaciers can also move big rocks to other locations. Another way glaciers change Earth is by creating lakes. For instance, Mirror Lake and the Great Lakes in the United States were formed by glaciers. Lake Louise in Canada is another example. Earth would look very different without the work of glaciers.

Extensive scaffolding activities teach students the **structure of academic writing.**

B Highlight the topic sentence.

C Highlight the sentences that provide major support for the topic sentence. Use a second color.

D Highlight the sentences that provide minor (or indirect) support for the topic sentence. Use a third color.

E Notice that the last sentence in the paragraph is the concluding sentence.

F Answer *T* for true or *F* for false. Compare answers with a partner.

____ **1.** The sentence "One way is by shaping the land" directly supports the main idea.
____ **2.** The paragraph gives examples to prove the major supporting sentences.
____ **3.** The main idea of the paragraph is that glaciers can carve the land.

5 Writing topic sentences and supporting sentences W

A Look at the topics. Write a topic sentence for each topic. Remember: Topic sentences state main ideas that can be proved. Examples, illustrations, facts, or statistics show or prove your point.

Example:
Topic: Glaciers and climate
Topic sentence: *Glaciers can tell us about climate.*
Major support: *For example, glaciers can show signs of global warming.*
Minor support: *Glaciers that melt fast can show that Earth is getting hot.*

1. Topic: Lakes and sources of freshwater
   Topic sentence: ________________
   Major support: ________________
   Minor support: ________________
2. Topic: The importance of glaciers
   Topic sentence: ________________
   Major support: ________________
   Minor support: ________________

B Go back to Step A. Write one major supporting sentence and one minor supporting sentence for each topic.

Students learn **key writing skills** such as summarizing and avoiding plagiarism. This early focus **prepares students** for later extended writing tasks.

Immersive Skill Building

The full-color **design mirrors university textbooks**, ensuring that students not only practice reading authentic texts, but also receive an **authentic university experience.**

Reading 1

THE WATER CYCLE

We call our planet *Earth*, but many people say that we should call it *Water*. Water covers more than 70 percent of our planet. Water is essential to life on Earth. We drink it, swim in it, clean with it, and use it in many other ways. Surprisingly, the amount of water on Earth 5 does not decrease even though we use so much of it every day. This is because nature recycles water in a process called *the water cycle* (also called *the hydrologic cycle*). The water cycle is the movement of water from Earth into the atmosphere and back to Earth again.

What are the steps of the water cycle?

Evaporation is the first step in the water cycle. This is the process that 10 changes water from a liquid to a gas. Energy from the sun produces **evaporation**. When the sun heats water, some of the water turns into a gas called *water vapor*. Water evaporates anywhere there is sun and water. Most evaporation of water on Earth is from the oceans, but there is also evaporation from lakes, rivers, and even from wet skin 15 and clothing.

evaporation the process that changes a heated liquid to a gas

PREPARING TO READ

Increasing reading speed

Academic classes often require a lot of reading. However, there is not always time to read every text slowly and carefully. Reading speed can be as important as reading comprehension. Here are some strategies for increasing your reading speed:

- Read the text straight through. Do not go back to any parts of it.
- Do not stop to look up words in a dictionary.
- Skip over words you do not know if they do not seem important.
- Try to guess the meaning of words that seem important.
- Slow down a little to understand important parts, such as definitions and main ideas.

A Read the text "Glaciers." Use the strategies listed above. For this task, do not read the boxed text on page 70.

1. Before you begin, fill in your starting time.
2. Fill in the time you finished.

Starting time: ______
Finishing time: ______

B Calculate your reading speed:

Number of words in the text (418) ÷ Number of minutes it took you to read the text = your reading speed

Reading speed: ______

Your reading speed = the number of words you can read per minute.

C Check your reading comprehension. Circle the correct answers. Do not look at the text.

1. Glaciers are made of
a. freshwater and saltwater.
b. soil and rocks.
c. layers of ice.

2. Glaciers can move
a. rocks and soil.
b. people and animals.
c. rivers and oceans.

3. Glaciers move
a. quickly.
b. slowly.
c. not at all.

Throughout each unit, **explanatory boxes describe each skill** and help **students understand why it is important.**

Academic Vocabulary and Writing

Chapter 3 Academic Vocabulary Review

The following words appear in the readings in Chapter 3. They all come from the Academic Word List, a list of words that researchers have discovered occur frequently in many different types of academic texts. For a complete list of all the Academic Word List words in this chapter and in all the readings in this book, see Appendix xx on page 00.

accessible	energy	generations	occur
approximately	environment	global	percent
contrast	environmentalists	labels (n)	source

Complete the following sentences with words from the list above.

1. In 2012, about half of the people in the United States drank soda every day, and 64 ________ drank coffee every day.
2. Air and water pollution is a ________ problem. People from many different countries struggle with this issue.
3. Some farmers use wind ________ to pump water from the ground.
4. Chewang Norphel's family has lived in India for many ________.
5. Oceans and rivers are a major ________ of income for people who fish their waters.
6. Most dolphins live in the ocean, but the pink river dolphin lives in a freshwater ________.
7. Mount Kosciuszko in Australia is considered an ________ mountain, and more than 100,000 people climb it every year.
8. Many people enjoy hiking in the mountains, but problems can ________. Do not hike in the mountains unless you are familiar with the area.
9. ________ often warn us that we might destroy our planet. They say we must not pollute the water, land, and air.
10. Today, glaciers cover about 10 percent of the land on Earth. This is a huge ________ to twenty thousand years ago; then, glaciers covered 32 percent of the land on Earth.

74 Unit 2 *Water on Earth*

Academic vocabulary development is **critical to student success.** Each unit includes **intensive vocabulary practice,** including words from the Academic Word List.

NOW WRITE

A Now write the first draft of your paragraph.

B Start with a clear topic sentence. Include at least three major supporting details. Use examples and statistics to explain and support your ideas. Make sure you add specific supporting examples and statistics. Use this checklist:

Are you including:

____ a topic sentence that states the main idea of the paragraph
____ major supporting details
____ minor details that illustrate the major or key support
____ a concluding sentence that restates the main idea (Be sure to make the concluding sentence a little different from the topic sentence.)
____ correct paragraph form and structure
____ vocabulary you learned in this chapter
____ correct sentences with subjects and verbs that agree

C Give your paragraph a title.

AFTER YOU WRITE

A Exchange paragraphs with a partner and read each other's work. Then discuss the following questions about both paragraphs:

- What is the most interesting information in your partner's paragraph?
- Does your partner's paragraph have correct form and structure? How do you know? Explain.
- Do the topic sentences make a claim about the topic?
- Are supporting details included, such as examples or facts?
- Are all of the ideas presented in a logical order? What are the transition words?
- Is all of the information on topic?
- Are there spelling or grammar mistakes?

B Think about any changes to your paragraph that would improve it. Then write a second draft of the paragraph.

150 Unit 3 *The Air Around Us*

Students complete each unit by **applying their skills** and knowledge in an extended writing task that **replicates university coursework.**

To the student

Welcome to *Academic Encounters 1 Reading and Writing: The Natural World!*

The *Academic Encounters* series gets its name because in this series you will encounter, or meet, the kinds of *academic* texts (lectures and readings), *academic* language (grammar and vocabulary), and *academic* tasks (taking tests, writing papers, and giving presentations) that you will encounter when you study an academic subject area in English. The goal of the series, therefore, is to prepare you for that encounter.

The approach of *Academic Encounters 1 Reading and Writing: The Natural World,* may be different from what you are used to in your English studies. In this book, you are asked to study an academic subject area and be responsible for learning that information, in the same way as you might study in a college or university course. You will find that as you study this information, you will at the same time improve your English language proficiency and develop the skills that you will need to be successful when you come to study in your own academic subject area in English.

In *Academic Encounters 1 Reading and Writing: The Natural World*, for example, you will learn:

- how to read academic texts
- ways to think critically about what you read
- how to write in an academic style
- methods of preparing for tests
- strategies for dealing with new vocabulary
- note-taking and study techniques

This course is designed to help you study in English in *any* subject matter. However, because during the study of this book, you will learn a lot of new information about research findings and theories in the field of sociology, you may feel that by the end you have enough background information to one day take and be successful in an introductory course in sociology in English.

We certainly hope that you find *Academic Encounters 1 Reading and Writing: The Natural World* useful. We also hope that you will find it to be enjoyable. It is important to remember that the most successful learning takes place when you enjoy what you are studying and find it interesting.

Author's acknowledgments

While not as all-consuming as writing the first edition, working on the second edition of *The Natural World* allowed me to revisit the process of creating something good. It also allowed me, once again, to enjoy the collaborative effort that goes into producing a textbook. I would like to express my sincere thanks to all the people at Cambridge who worked so hard to ensure that the best book possible would be published, and a special thanks to Bernard Seal, Series Editor, Christopher Sol Cruz, Editorial Manager, and Susan Johnson, Senior Development Editor. I would also like to acknowledge the people whose support and guidance made the book possible the first time around: Kathleen O'Reilly, Amy Cooper, Donna Prather, Patty Reiss, and Beth Edwards. And mahalo nui loa to Yoneko Kanaoka, for the many years of friendship and teaching adventures we have shared.

I am also very grateful to Caitlin Mara, Managing Editor, Robin Berenbaum, Associate Editor, and all of those who made invaluable comments and suggestions.

My students have taught me many things over the years. Like most teachers, I have learned a lot in the classroom, and I have tried to remember those lessons as I wrote this book.

Finally, I wish to thank my family and friends on the East Coast and in Hawaii, especially my husband, Patrick, and my daughters, Emma and Fiona, who remind me each day of what matters most.

Jennifer Wharton

Publisher's acknowledgments

The first edition of *Academic Encounters* has been used by many teachers in many institutions all around the world. Over the years, countless instructors have passed on feedback about the series, all of which has proven invaluable in helping to direct the vision for the second edition. More formally, a number of reviewers also provided us with a detailed analysis of the series, and we are especially grateful for their insights. We would therefore like to extend particular thanks to the following instructors:

Doreen Ewert, Indiana University, Bloomington, Indiana
Monique Grindell, Pacific University, Forest Grove, Oregon
Anne Lech, Northwest Missouri State University, Maryville, Missouri
Ursala McCormick, Lewis & Clark College, Portland, Oregon

Unit 1
Planet Earth

In this unit, you will look at the physical features of our planet and the different ways Earth grows and changes. In Chapter 1, you will see how Earth is a part of the universe. You will also discuss the ways in which our planet is unique, or different from, all the others. In Chapter 2, you will focus on various processes that help create and shape our planet.

Contents

In Unit 1, you will read and write about the following topics.

Chapter 1 The Physical Earth	**Chapter 2** The Dynamic Earth
Reading 1 Our Solar System	**Reading 1** Plate Tectonics
Reading 2 Earth's Four Systems	**Reading 2** Volcanoes
Reading 3 Rocks on Our Planet	**Reading 3** Earthquakes

Skills

In Unit 1, you will practice the following skills.

 Reading Skills	 Writing Skills
Thinking about the topic Previewing art Asking and answering questions about a text Previewing key parts of a text Using headings to remember main ideas Building background knowledge Reading boxed texts Illustrating main ideas Reading for main ideas	Parts of speech Comparative adjectives Writing complete sentences Writing simple and compound sentences Writing definitions Pronoun reference Showing contrast Using correct paragraph form Using correct paragraph structure Writing first drafts
 **Vocabulary Skills**	 **Academic Success Skills**
Words from Latin and Greek Cues for finding word meaning Learning verbs with their prepositions Previewing key words Prefixes Prepositional phrases Using grammar, context, and background knowledge to guess meaning	Highlighting Making a pie chart Answering multiple-choice questions Labeling diagrams Reading maps Answering true/false questions

Learning Outcomes

Write an academic paragraph about a place on Earth you like

Previewing the Unit

Previewing means looking at one thing before another. It is a good idea to preview your reading assignments. Read the contents page of every new unit. Think about the topics of the chapters. You will get a general idea of how the unit is organized and what it is going to be about.

Read the contents page for Unit 1 on page 2 and do the following activities.

Chapter 1: The Physical Earth

A How much do you know about our solar system and planet Earth? Look at the photographs. Then answer the questions below.

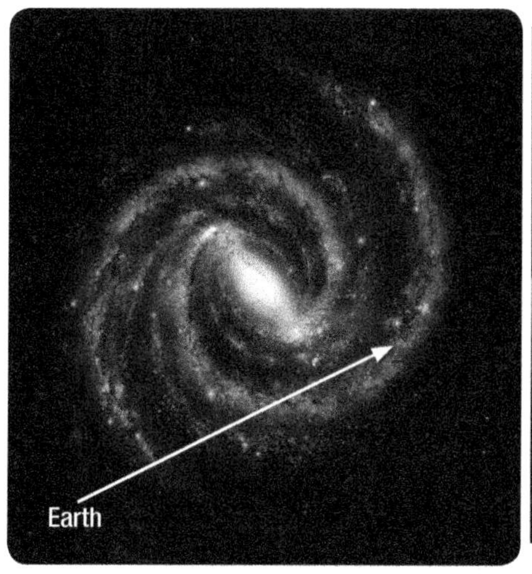

1. Earth is only one of the planets in the universe, but it is very special. What makes Earth unique, or different from, all the other planets?
2. What makes it possible for people to live on Earth? What does Earth provide so that we are able to live here?

B Compare your answers in a small group.

Chapter 2: The Dynamic Earth

A Discuss the following questions in a small group.

1. Earth is always moving in different ways. How does Earth move? Did you ever feel it move? Describe your experience.
2. Why do you think Earth moves?
3. Earth's surface is not flat. It has many natural features such as mountains and lakes. Make a list of the natural features on Earth. Choose two features from the list and answer this question: How was each feature created?

B Share your answers with the class.

Chapter 1
The Physical Earth

PREPARING TO READ

1 Thinking about the topic Ⓡ

Thinking about the topic of a reading before you read can make a text easier to understand.

The text you are going to read is about the solar system. It discusses some of the objects we see in the sky. How much do you know about these objects? Discuss the following questions in a small group.

1. What are some things that you see in the sky during the day?
2. What are some things that you see in the sky at night?
3. Would you like to be a scientist who studies the sky? Why or why not?

2 Previewing art Ⓡ

Art in textbooks illustrates ideas in a text. Previewing photographs and illustrations can help you quickly grasp these ideas.

A Look at the illustration of our solar system on page 5. Then discuss the following questions with a partner.

1. How many planets are there in the solar system? Do you know any of their names?
2. In what ways are the planets different from one another?
3. Can you find our planet, Earth? If you can, draw an arrow (➔) to it.
4. What is at the center of our solar system? Is it a planet?

B Look at the photograph at the top of page 5. Then discuss the following questions with a partner.

1. Who is the person in the photograph?
2. What is the name of the instrument with him?
3. What do these instruments do? Did you ever look through one? Describe your experience.

Clyde Tombaugh

Reading 1

OUR SOLAR SYSTEM

Our home in the universe is planet Earth. It is one of eight planets that **orbit**, or circle, the sun. The sun is a star, that is, a giant ball of hot gases. It is the center of our **solar system**. There are billions of other stars in the sky, but the sun is the star closest to Earth. Our solar system also includes moons, which orbit planets. The moon we see in the night sky orbits Earth.

orbit travel in a circle around a larger object

solar system the sun and the planets that move around it

We usually list our solar system's planets in order of their distance from the sun: Mercury, Venus, Earth, Mars, Jupiter, Saturn, Uranus, and Neptune. We can divide the planets into two groups: terrestrial planets and gas giant planets.

Terrestrial, or Earthlike, planets have solid, rocky surfaces. Mercury, Venus, Earth, and Mars are **terrestrial planets**. Earth is the only planet that has large amounts of liquid water, and it is the only planet that has life. Astronomers (scientists who study the stars and planets) believe that a long time ago, Mars had rivers and oceans, just like Earth, but that now all the water is either frozen or underground.

terrestrial planet a planet with a solid, rocky surface

Gas giant planets are much larger than terrestrial planets. All **gas giant planets** are made of gases, not solid rock. These planets have rings around them. The rings are made of tiny pieces of rock, dust, or ice. Jupiter, Saturn, Uranus, and Neptune are gas giant planets. Jupiter is the largest planet. It is about a thousand times bigger than Earth.

gas giant planet a planet made of gases, not solid rock

Outside of our solar system, there are billions of other stars. Astronomers now know that some of these stars have planets, and the planets orbit these stars. This means that there are other solar systems in the universe in addition to our own. Perhaps we will even find another planet with life on it someday.

The Story of Pluto

Are there eight or nine planets in our solar system? Before 1930, everyone thought there were eight. Then in 1930, a young man surprised the world. Clyde Tombaugh was 24 years old. He was not a professional astronomer. He even made his own telescope. He looked through his telescope and discovered a new planet in the solar system. He discovered Pluto. Pluto is very small, smaller than Earth's moon. It is also farther from the sun than the other planets, so it is colder and darker. After Tombaugh's discovery, people accepted it as fact that our solar system has nine planets.

Then, in 2006, the International Astronomical Union (IAU) changed the definition of "planet." According to the new definition, a planet has to have certain features. For example, it has to make a circular orbit around the sun. It also has to be big enough and strong enough to move objects in its path away. Pluto differs from the other planets in both these ways. First, it does not have a circular orbit, and it even crosses Neptune's orbit. In addition, it is not able to move objects out of its way as it orbits the sun. For these reasons, the IAU reclassified Pluto from a planet to a plutoid. So now scientists tell us once again that there are only eight planets.

AFTER YOU READ

1 Asking and answering questions about a text (R)

Asking and answering questions about a text checks your understanding.
When you do this, you discover what you know and do not know about the reading.
You can do this alone or with a partner.

Work with a partner to complete the following activities.

A Reread paragraph 1 of the text "Our Solar System." Use the questions below to ask and answer questions about the text. Add at least two more questions to the list.

1. How many planets are there in our solar system?
2. What does *orbit* mean?
3. Is the sun a planet or a star?
4. (Add your own question.)
5. (Add your own question.)

B Now reread paragraph 2 of the text. Write two or three questions about the paragraph. Then take turns asking and answering the questions.

C Reread the rest of the text and the boxed text "The Story of Pluto." Continue asking and answering questions with your partner.

2 Words from Latin and Greek (V)

Many English words, or parts of words, come from other languages. Words that come from Latin and Greek are especially common in science. Look at these examples:

terr- means "earth" or "land"

sol- means "sun"

astro- means "star"

Learning Greek and Latin word parts helps you put new words together.
You can then quickly guess the basic meanings of many new words in science texts.

A Go back to the text "Our Solar System" and the boxed text "The Story of Pluto." Find words that start with *terr-*, *sol-*, or *astro-* and put a check mark (✓) above them.

B Start a chart of word parts from Latin and Greek in your notebook. Follow this model:

Word part from Latin or Greek	Meaning	English example	Meaning
terr-	earth, land	terrestrial	relating to Earth

3 Cues for finding word meaning V

Academic texts will have many new words. However, when you read a text, do not stop to look up every unfamiliar word. The definition of a new word is often in the text. Learn to look for cues to its meaning.

Look at these sentences from "Our Solar System":

- It is one of eight planets that orbit, or circle, the sun.
- The sun is a star, that is, a giant ball of hot gases.
- Terrestrial, or Earthlike, planets have solid, rocky surfaces.
- Astronomers (scientists who study the stars and planets) believe that a long time ago, Mars had rivers and oceans, just like Earth.

Notice that the sentences use three cues to present definitions: *or*, *that is*, and parentheses ( ).

A Read the sentences below. Notice the key words in **bold** and the definitions. Find the cues that signal the definitions and circle them.

1. In ancient times, sailors often used **constellations** (groups of stars that form imaginary pictures and have names) to safely find their way across the ocean.
2. Astronomers have found more than 750 **extrasolar** planets, that is, planets outside of our solar system.
3. A **supernova**, or extremely bright explosion of a star, is a very rare occurrence.
4. The name of our **galaxy** (a group of stars, gas, and dust held together by gravity) is "the Milky Way."

B Write the key word from each sentence in Step A and its definition.

1. ________________ :
   ____
2. ________________ :
   ____
3. ________________ :
   ____
4. ________________ :
   ____

C Write three sentences with definitions of the following terms. Use *or*, *that is*, or parentheses ( ). Be sure to use correct punctuation. Write on a separate piece of paper.

1. telescope: *A telescope (an instrument that makes faraway objects look larger) is an important tool for an astronomer.*
2. solar system:
3. Mercury or Pluto:
4. gas giant planet:

4 Parts of speech W R V

Words are also parts of speech. Parts of speech say whether a word names, does, or describes. Understanding the parts of speech in English helps your writing and reading. You can write clearer, more logical sentences and better understand what you read. In this chapter, we review three parts of speech:

A noun names a person, place, thing, or idea (*planet*, *moon*, *belief*).

A verb does an action or is a state of being (*orbit*, *be*, *have*).

An adjective describes a noun or pronoun (*rocky*, *giant*, *hot*, *it*).

A Reread paragraph 1 of the text "Our Solar System." Then do the following.

1. Underline all of the nouns.
2. Draw two lines under all of the verbs.
3. Circle all of the adjectives.

Compare answers with a partner.

B Read the paragraph below. Label each noun (**n**), verb (**v**), and adjective (**adj**).

Mars is an interesting planet. In some ways, it is similar to Earth. It has weather and seasons. It also has canyons and mountains. However, Mars is a very different planet from Earth. It is much smaller than Earth, and it is much colder. In addition, there are no people on Mars.

C Read these sentences. There is a missing word in each one. Decide what part of speech is missing and write it in the first blank.

__*verb*__ **1.** Pluto ________ two small moons.

______ **2.** Mercury is a ________ planet.

______ **3.** Saturn has many beautiful ________ .

______ **4.** Earth has one ________ .

______ **5.** Some people ________ that there is life on other planets.

D Now complete each sentence in Step C. Use an appropriate word. Use the part of speech to help you. Compare your sentences with a partner's.

Example: __*verb*__ Pluto __*has*__ two small moons.

5 Comparative adjectives W V

A **comparative adjective** shows the difference between two people, places, or things. Sometimes a comparison includes a group of people, places, or things. Look at these examples:

- Tombaugh was **younger** than most other astronomers when he discovered Pluto.
- Jupiter is **farther** from the sun than Earth.
- Saturn's rings are **more beautiful** than Jupiter's rings.
- Venus is **hotter** than the other planets in our solar system.

To form comparative adjectives, follow these guidelines:

For **one-syllable** adjectives, add *-er*. If the adjective ends in *-e*, add only *-r*.

small → smaller
dark → darker
close → closer

For **one-syllable adjectives that end with a single vowel and a consonant**, double the final consonant and add *-er*.

hot → hotter
big → bigger
red → redder

For **adjectives with two or more syllables**, add *more* before the adjective. If the adjective ends in *-y*, change the *-y* to *-i* and add *-er*.

important → more important
solid → more solid
happy → happier

Irregular adjectives do not follow patterns. Check your dictionary for a complete list.

good → better
bad → worse
far → farther

To compare two nouns in the same sentence, use *than* after the comparative adjective and before the second noun.

noun 1		comparative adjective		noun 2
Earth	is	smaller	**than**	Jupiter.

A Go back to paragraph 4 of "Our Solar System" and to the boxed text "The Story of Pluto." Find the comparative adjectives. Underline them. How many did you find?

B Write the comparative form of each adjective.

1. dark ________
2. hot ________
3. solid ________
4. icy ________
5. small ________
6. big ________
7. strong ________
8. rocky ________

C Complete each sentence with a comparative adjective. Choose an adjective from the box. Use each word once. Use correct forms and add than. Be sure that each sentence is true, based on the information in the texts.

close	cold	far	hot	large	rocky

1. Uranus is farther from the sun than Mercury.
2. Pluto is ________ ________ Venus.
3. Mars is ________ ________ Pluto.
4. Earth is ________ to the sun ________ Neptune.
5. Mercury is ________ ________ Saturn.
6. Jupiter is ________ ________ Uranus.

D Compare Jupiter and Pluto. Write three or four sentences. Use comparative adjectives with *than*.

Example: Jupiter is closer to the sun than Pluto.

1. ____
2. ____
3. ____
4. ____

PREPARING TO READ

Previewing key parts of a text Ⓡ

Previewing key parts of a text tells you the main ideas of the text. To preview key parts, look carefully at the title, the introduction, and the headings. It is also a good idea to read the first sentence of each paragraph.

A Read these key parts of the text "Earth's Four Systems" on page 13.

- the title
- the short introductory paragraph at the beginning of the text
- the headings
- the first sentence of each paragraph

B Answer the following questions with a partner.

1. How many systems does Earth have?
2. What are their names?

C Now complete the chart.

1. Write the names of Earth's systems in the first column.
2. Then match the following key features to the systems. Write the feature next to the appropriate system:

living things
water
Earth's crust and the top layer of the mantle
air

Name of the system	Key feature(s)
lithosphere	Earth's crust and the top layer of the mantle

Reading 2

EARTH'S FOUR SYSTEMS

Think about Earth from the point of view of an astronaut. From outer space, Earth looks like one solid blue ball. In fact, our planet is much more complex. It is actually made up of four very different, but interconnected, systems: the lithosphere, the hydrosphere, the atmosphere, and the biosphere.

The lithosphere

The lithosphere is the hard surface of Earth. It has two layers. The first layer is the **crust**. The crust is a thin layer of rock that covers the whole planet. Its thickness ranges from about 5 to 80 kilometers. The second layer is called the mantle. The mantle is directly under the crust. The lithosphere is not one solid piece of rock. It is made up of many smaller pieces called plates.

crust Earth's hard outer layer

The hydrosphere

The hydrosphere is all the water on Earth, including oceans, lakes, rivers, **glaciers**, rain, and snow. Water covers more than 70 percent of Earth. Approximately 97 percent of Earth's water is saltwater from oceans, and 3 percent is freshwater from glaciers, lakes, rivers, and groundwater (water under the ground).

glacier a very large amount of ice that moves slowly over land

The atmosphere

The atmosphere is the air surrounding Earth. It is made up mostly of gases. The primary gases are nitrogen and oxygen. Gases in the atmosphere create air for us to breathe, and they protect Earth from the sun's **ultraviolet radiation**. The atmosphere is also where weather conditions, such as clouds and storms, form.

ultraviolet radiation a form of energy that comes from the sun in rays, or lines, that we cannot see

The biosphere

The biosphere is made up of all the living things on Earth. It includes humans, animals, and plants. Life on Earth is very diverse, but all living things share certain features. For example, they all eat, breathe, and grow.

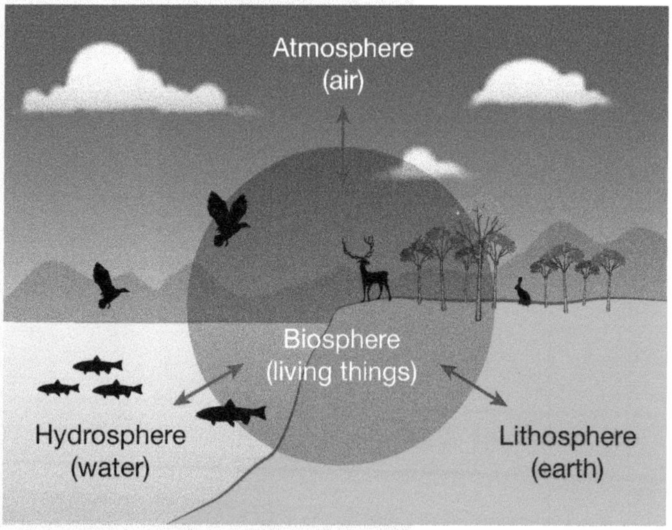

Figure 1.1 Earth's four systems

The interconnections of Earth's systems

The lithosphere, the hydrosphere, the atmosphere, and the biosphere connect with each other in important ways. We humans are part of the biosphere, but we live on the lithosphere. We depend on the atmosphere for air to breathe and on the hydrosphere for water to drink. In fact, these connections are so strong that a change in one system can affect the others. Consider this example: Driving a car contributes to air pollution in the atmosphere. Air pollution causes Earth to grow warmer. Warmer temperatures cause important changes in the hydrosphere: Glaciers melt and ocean levels rise. These changes to the hydrosphere affect the humans, animals, and plants of the biosphere. For example, people who live in coastal areas along the ocean are in danger of losing their homes because of **floods**. The polar bear is gradually losing its **natural habitat** because of warming temperatures in the Arctic. When you think about these interconnections among the systems, it is easy to see that our planet is very complex.

flood a large amount of water covering an area that is usually dry

natural habitat the place where an animal or a plant usually lives

AFTER YOU READ

1 Highlighting A V R

Highlight important information in a text. This is one way to remember what you read. Mark key words and ideas in colors that are easy to see. Remember to highlight only the important information. A text that has too many highlighted sentences is not useful.

A Go back to the text "Earth's Four Systems." Highlight the names of Earth's systems and the most important feature of each one.

B Go back to the text again. Find an explanation of how humans are connected to Earth's four systems. Use a different color and highlight the explanation.

C Compare your work with a partner's.

2 Words from Latin and Greek V

Remember that many words in science come from Latin or Greek. Knowing the meaning of all the parts can help you better grasp the exact meaning of the word. For example, *hemisphere* is made up of *hemi-* (half) and *sphere* (ball). Earth is shaped like a ball. The word *hemisphere* means "half of Earth."

A Match the word parts with the meanings. Use the information in the reading.

____	**1.** litho-	**a.** water
____	**2.** hydro-	**b.** life
____	**3.** atmo-	**c.** rock, stone
____	**4.** bio-	**d.** gas, vapor

B Circle the correct word in each sentence.

1. A spherical cloud is shaped like a (*square / circle / diamond*).
2. Lithology is the study of the physical qualities of (*rocks / water / gases*).
3. Countries that use hydropower to create energy are using (*water / rocks / air*).
4. Atmospherology is the study of (*water / rocks / gases*).
5. A biologist works with (*rocks / gases / living things*).

C Look at the sentences in Step B. Circle two words with *-logy*.

What do you think the word part *-logy* means? Write a short definition.

____

Then add *-logy* to your chart of word parts from Latin and Greek.

D Add the new word parts from the box above and from step A on page 15 to your chart of word parts from Latin and Greek. Include the example words given in the box and in Step B. (Review "Words from Latin and Greek" on page 17 if necessary.)

3 Learning verbs with their prepositions V

Some verbs often occur with specific prepositions, for example: *benefit from*, *pay for*, *think about*. When you learn a new verb, notice if a preposition follows it. Try to learn the verb and the preposition together as a unit.

A The verbs in **bold** frequently occur with specific prepositions. Complete the sentences with the correct prepositions. Find the verbs in the reading, if needed.

1. Its thickness **ranges** __________ about 5 to 80 kilometers. (Par. 2)
2. Gases in the atmosphere create air for us to breathe, and they **protect** Earth __________ the sun's ultraviolet radiation. (Par. 4)
3. The lithosphere, the hydrosphere, the atmosphere, and the biosphere **connect** __________ each other in important ways. (Par. 6)
4. We **depend** __________ the atmosphere for air to breathe and on the hydrosphere for water to drink. (Par. 6)
5. Driving a car **contributes** __________ air pollution in the atmosphere. (Par. 6)

B Complete the sentences. First, add the correct preposition for each verb. Then choose an appropriate ending from the box.

the sun's dangerous rays	-238°C to -228°C	good health
other scientists all over the world	a cold environment	

1. Scientists use the Internet to connect ___with___ ___other scientists all over the world___ .
2. Sunscreen and sunglasses protect people __________ ____________________ .
3. The temperature on Pluto ranges __________ ____________________ .
4. Polar bears depend __________ ____________________ .
5. Drinking clean water and breathing clean air contribute __________ ____________________ .

C Complete these sentences with something that is true for you.

1. I depend ______________________________.
2. The summer temperature where I live ranges ______________________________.
3. I want to contribute ______________________________.

4 Making a pie chart Ⓐ Ⓡ

Textbooks often use statistics, that is, a group of facts that are stated as numbers. Sometimes it is helpful to organize a group of statistics in a pie chart. A pie chart shows the parts of a whole. Pie charts make the statistical information easier to read and understand.

A Read the following sentence from the text "Earth's Four Systems." Notice how the pie chart on the right organizes the information.

Water covers more than 70 percent of Earth.

Water and Land on Earth

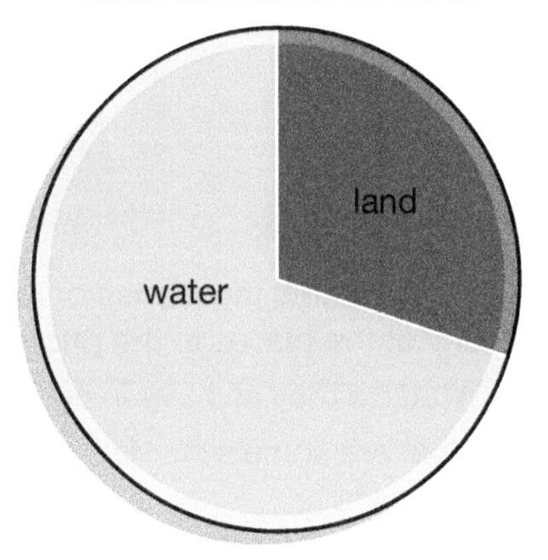

B Reread paragraph 3 of the text. On a separate piece of paper, make a pie chart. Represent the amount of freshwater and saltwater on Earth in your chart. Look at your chart and write two sentences about the information. You can start this way:

Approximately ________ percent of the water on Earth is freshwater.

C Work in pairs or small groups. Survey your classmates and record their answers. You can use one of the questions below or ask your own question about our planet.

- *Do you think there is life beyond Earth?*
- *Which of Earth's four systems are you most interested in?*

D Make a pie chart to organize the results of your survey. Give the chart a title and write a few sentences about the information. For example, you might write sentences like these:

In our class, 75 percent of the students think there is life on other planets.

In our class, 40 percent of the students are most interested in learning more about the hydrosphere.

E Present your pie chart and information to the class.

PREPARING TO READ

1 Thinking about the topic

Look at these photographs. Then discuss the questions below in a small group.

a.

b.

c.

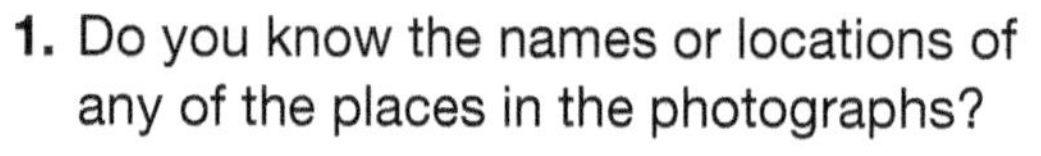

1. Do you know the names or locations of any of the places in the photographs?
2. What do they all have in common?
3. What are some other famous places made of rock?
4. What do people use rocks for? Try to think of at least three uses.
5. Did builders use any rocks to construct your school or your home? Are there any things inside your school or home that are made of rock?

2 Previewing key parts of a text

A Read these key parts of the text "Rocks on Our Planet" on pages 19–20:

- the title
- the introductory paragraphs
- the headings
- the photographs and illustrations

B Answer these questions with a partner.

1. How many main types of rocks does Earth have?
2. What are their names?
3. What is the rock cycle?

Reading 3

ROCKS ON OUR PLANET

Earth is a terrestrial planet, that is, a planet with a rocky surface. It is covered with rocks of all ages. The oldest rocks in Earth's crust are more than three billion years old. The youngest ones are just a few minutes old. All rocks are made of minerals, or inorganic (nonliving) matter.

5 Different types of rocks form in different ways, but all rocks come from the same original hot material, **magma**, deep inside Earth. The three main types of rocks are igneous, sedimentary, and metamorphic.

magma very hot melted, or liquid, rock that is deep inside Earth

Main types of rocks

Igneous means "relating to fire." When the hot, fiery magma rises up through Earth's crust, it cools and becomes igneous rock. Sometimes 10 the melted rock cools under the surface of Earth, but sometimes magma erupts from a volcano as **lava** and cools on Earth's surface. Granite, basalt, and pumice are examples of igneous rocks.

lava hot, melted rock that flows from a volcano

Rocks are very strong, but wind and rain over time can break off tiny pieces. These pieces of rock often end up at the bottom of a river 15 or ocean. This layer of little rocks is called *sediment*. After thousands of years, many layers of sediment form on top of each other. The weight of all of the layers presses the sediment tightly together, and it becomes solid rock. Some common sedimentary rocks are limestone, sandstone, and shale.

20 The heat and pressure deep inside Earth can actually change one type of rock into another. This process is called *metamorphosis*, or the process of changing one thing into another. Rocks that form in this way are called *metamorphic rocks*. For example, great heat and pressure over a long time can change limestone, a sedimentary rock, 25 into marble, a metamorphic rock.

The rock cycle

Over time, any type of rock can change into any other type. This process is called *the rock cycle*. Magma cools and forms igneous rocks. Igneous rocks 30 break into small pieces and form

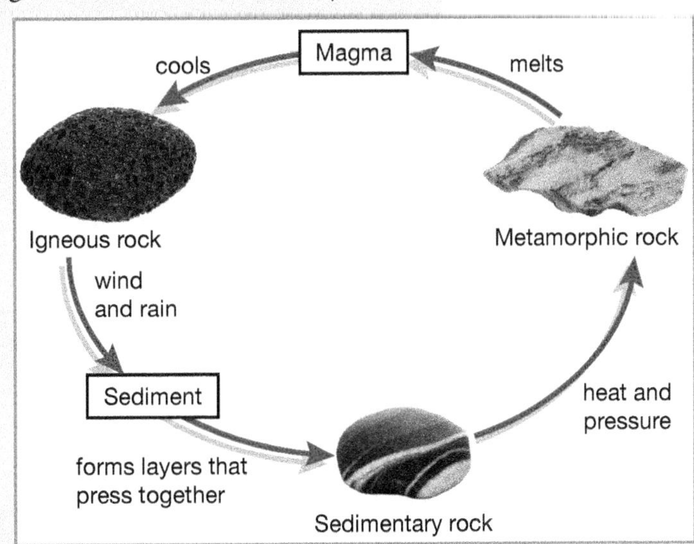

Figure 1.2 The rock cycle

sedimentary rocks. Deep inside Earth, great heat and pressure act on sedimentary and igneous rocks and change them into metamorphic rocks. When all three kinds of rock move even deeper into Earth, they melt and become magma, and the rock cycle begins all over again. In this way, for millions of years, the rocks in Earth's crust have continuously changed form. Rocks are natural recyclers.

Save the Rocks!

There is something very unusual in Narendra Luther's living room – a giant, two-billion-year-old rock. The rock goes all the way up through the ceiling to the second floor. This rock is just one of many that formed in Hyderabad, India, billions of years ago. The people gave names to some of the rocks, such as Bear's Nose and Stone Heart. They used many of them to make temples and some to make billboards. They also had to destroy a large number of the rocks to make room for new buildings such as offices, apartments, hotels, and shopping malls. As the city develops, there is less and less room for these giant reminders of Earth's past.

Many people in Hyderabad want to save some of the rocks. The city now has a rock park, and several new buildings include rocks as part of their design, just like Mr. Luther's house. In this way, people can enjoy the new things that come with development, and they can also save part of their past.

AFTER YOU READ

1 Answering multiple-choice questions A R

One common type of question on tests and in textbooks is the multiple-choice question. Here are some strategies for answering this type of question:

- Read the question several times. Make sure you understand it before you try to choose an answer.
- Think of the correct answer and then look for it in the choices.
- Read all of the choices before you make a decision. Do not stop reading as soon as you think you have found the correct answer.

Answer the questions below based on the reading "Rocks on Our Planet." Then compare your answers with a partner's.

1. What are all rocks made of?

a. lava

b. water

c. fire

d. minerals

2. Which one of the following is a type of rock?

a. sedimentary

b. organic

c. terrestrial

d. metallic

3. Where do igneous rocks form?

a. on Earth's surface

b. under Earth's surface

c. in a river

d. both on and under Earth's surface

4. Where do many sedimentary rocks form?

a. on Earth's surface

b. under Earth's surface

c. in a river

d. in the wind

5. What forces create metamorphic rocks?

a. wind and rain

b. heat and pressure

c. cooling and melting

d. erupting and breaking

6. Which process causes rocks to change form continuously over time?

a. the formation of sediment

b. the rock cycle

c. volcanic eruptions

d. the cooling of magma

2 Labeling diagrams A R

Labeling diagrams with key words helps you understand and remember complex information in a text. Science texts frequently discuss processes and parts of things, so this strategy is especially helpful when you read science texts.

A Look at the diagrams below. Review the following terms in the reading and label them in the diagram: *lava*, *magma*, and *igneous rock*.

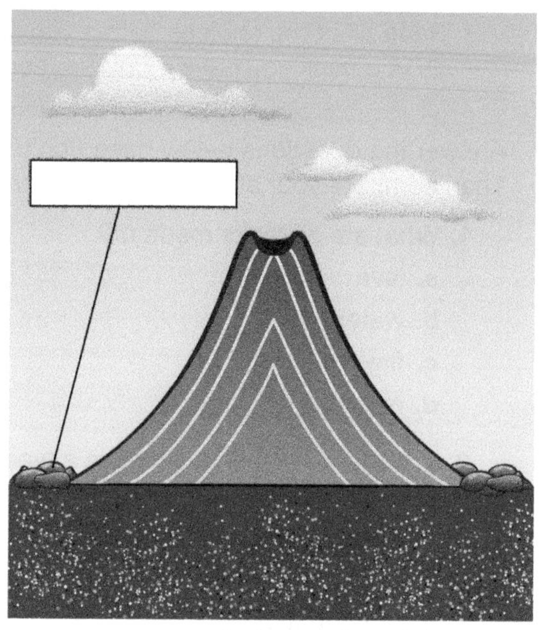

A volcano during and after eruption

B Look at the diagram of the rock cycle below. Label the blanks with these terms: *metamorphic rock*, *sedimentary rock*, *igneous rock*, *sediment*, and *magma*.

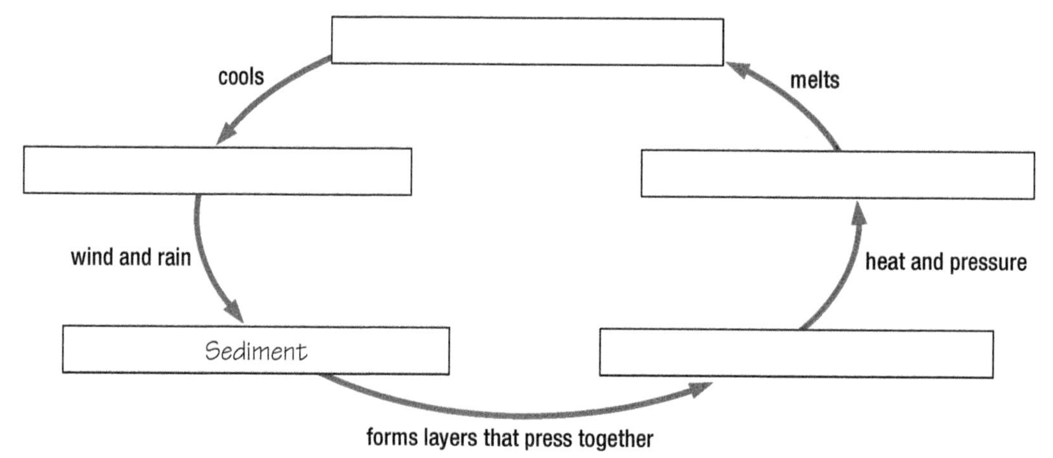

The rock cycle

C Look at Figure 1.2 on page 19 to check your work for Step B.

3 Cues for finding word meaning Ⓥ

A You learned that there may be cues to the meanings of words in a text.

1. These are three key terms in "Rocks on Our Planet." Find them in the text and circle them.
   - terrestrial planet
   - inorganic
   - metamorphosis
2. Underline the definition of each key term. Highlight the cue that helped you find the definition.
3. Compare your answers with a partner.

B Another cue to the meaning of new words is the phrase "X *is called* Y." Look at the example below. Notice the location of the key term *metamorphosis*. What is the meaning of *metamorphosis*?

The process of changing one thing into another is called metamorphosis.

C Reread the sections of the text below. Find and circle *sediment* and *the rock cycle*. Underline the cues to the meaning of these terms.

Rocks are very strong, but wind and rain over time can break off tiny pieces. These pieces of rock often end up at the bottom of a river or ocean. This layer of little rocks is called *sediment*. After thousands of years, many layers of sediment form on top of each other. The weight of all of the layers presses the sediment tightly together, and it becomes solid rock. Some common sedimentary rocks are limestone, sandstone, and shale.

Over time, any type of rock can change into any other type. This process is called *the rock cycle*. Magma cools and forms igneous rocks. Igneous rocks break into small pieces and form sedimentary rocks. Deep inside Earth, great heat and pressure act on sedimentary and igneous rocks and change them into metamorphic rocks. When all three kinds of rocks move even deeper into Earth, they melt and become magma, and the rock cycle begins all over again.

D Write a sentence for each term that explains the meaning of the term. You could begin your sentences this way:

Sediment is . . .

The rock cycle is . . .

Chapter 1 Academic Vocabulary Review

The following words appear in the readings in Chapter 1. They all come from the Academic Word List, a list of words that researchers have discovered occur frequently in many different types of academic texts. For a complete list of all the Academic Word List words in this chapter and in all the readings in this book, see the Appendix on page 206.

area	cycle	diverse	layer (n)	primary	professional
complex	design (n)	features	period	process (n)	ranges (v)

Complete the following sentences with words from the list.

1. A ________ marine biologist has studied at a university and has received a degree in marine science.
2. One of the unique ________ of the planet Mars is its reddish color.
3. The ________ of changing liquid water into ice requires a minimum temperature of 0° Celsius.
4. Earth is home to many ________ languages and cultures.
5. If an ________ is too cold, people may not want to live there.
6. Humans often have a ________ relationship with nature; they need it to survive but they don't always take care of it.
7. The ________ of a house is the overall plan of the house.
8. Similar to humans, stars are born, grow, exist, and then die; however, while most humans live less than 100 years, the life ________ of a star is billions of years.
9. The life span of a butterfly ________ from a few days to 12 months, depending on the species.
10. Two ________ uses of rocks are for building materials and decorative items.

Developing Writing Skills

In this section, you will learn to write complete sentences. Complete sentences are the building blocks of all academic writing. You will write eight to ten complete sentences. You will also use what you learn here to complete the writing assignment at the end of this unit.

Writing Complete Sentences

A sentence in English must have at least one subject. The subject of a sentence names who or what the sentence is about. It can be a single noun, a noun phrase (a noun + other words that give more information about the noun), or a pronoun. Pronouns are words that take the place of nouns. The subject pronouns are *I*, *you*, *he*, *she*, *it*, *we*, and *they*.

Common mistakes with subjects:

1. There is no subject in the sentence.

   INCORRECT: Discovered a new planet.

   To correct this sentence, add a subject.

   CORRECT: **An astronomer** discovered a new planet.

2. The same subject is written twice, usually both as a noun and as a pronoun. This is called a double subject. In English, sentences cannot have a double subject.

   INCORRECT: Earth it is a terrestrial planet.

   To correct this sentence, delete one of the subjects.

   CORRECT: **Earth** is a terrestrial planet.
   It is a terrestrial planet.

Sentences In English must also have at least one verb. Verbs are words of action (*erupt*, *press*, *move*) or being (*be*, *have*). Sometimes verbs are one word (*erupt*), sometimes they are two words (*is erupting*) or even three words (*has been erupting*).

Common mistakes with verbs:

1. There is no verb in the sentence.

   INCORRECT: Earth terrestrial planet.

   To correct this sentence, add a verb.

   CORRECT: Earth **is** a terrestrial planet.

2. There is only half of a verb. This often happens with the *-ing* form. A verb that ends in *-ing* cannot be a verb by itself. It needs a helping verb in order to be a complete verb.

   INCORRECT: The volcano erupting.

   To correct this sentence, add a helping verb or change the verb to a different form.

   CORRECT: The volcano **is** erupting.
   The volcano **erupted**.
   The volcano often **erupts**.

A Read the following paragraph. The subject of each sentence is underlined once and the verb is underlined twice.

> The lithosphere is the hard surface of Earth. It has two layers. The first layer is the crust. The crust is a thin layer of rock. It covers the whole planet. The thickness of the crust ranges from about 5 to 80 kilometers. The second layer is called the *mantle*. The mantle is directly under the crust. The lithosphere is not one solid piece of rock. Instead, it is made up of many smaller pieces. Scientists call these pieces plates.

B Now read this paragraph and underline the subject of each sentence once and the verb twice.

> The atmosphere is the air surrounding Earth. It is made up of gases. The primary gas is nitrogen. The gases in the atmosphere create air for us to breathe. They also protect Earth from the sun's ultraviolet radiation. Clouds form in the atmosphere. These clouds produce rain and snow.

C Each sentence in the paragraph below has one mistake. There are three mistakes with subjects and three mistakes with verbs. Mark the mistakes and correct them. Compare your answers with a partner. Then rewrite the paragraph on a separate piece of paper.

> Narendra Luther having something very unusual in his house. Is a giant, two-billion-year-old stone. This rock just one of many in the city of Hyderabad, India. The people in the city they named some of the rocks. Used many to make temples or billboards. People destroying other rocks to make room for new development.

D Now write eight to ten complete sentences about planet Earth. Imagine that someone from another planet is coming to visit Earth and that you are the guide. What do you want to tell the visitor about our planet? Follow the steps below.

1. Have a short conversation with a partner. Talk about the information you want to share with the visitor. Make a list of your ideas.
2. Now write sentences that express your ideas. Be sure to:
   a. Include information about the solar system, Earth's four systems, and rocks.
   b. Include new vocabulary from this chapter.
   c. Be sure that each sentence has a subject and a verb and that you use the correct parts of speech.
3. When you finish, exchange sentences with a partner and read each other's work.
   a. Look at the structure of the sentences. Do they all have a subject and a verb? Find and underline the subjects in your partner's sentences. Draw two lines under the verbs.
   b. Check (✓) your partner's three best sentences.
   c. Discuss the best sentences. Explain your reasons for your choices. Then, talk about any problems with subjects and verbs in your partner's sentences.
4. Now reread your own sentences. Make changes to improve your sentences and rewrite them.

Chapter 2
The Dynamic Earth

PREPARING TO READ

Previewing key words V

Learn the meaning of key words before you read. Key words are used to express important ideas in a text. Learning them before you read can make the text easier to understand.

A Read the following sentences. Think about the meanings of the words in **bold**. Use all the words in each sentence to help you.

_e_ **1.** Earth has seven **continents**. Asia is the largest continent, and Australia is the smallest.

____ **2.** The Pacific Ocean surrounds (goes all around) the **islands** of Hawaii.

____ **3.** The Pyrenees are mountains that form a **boundary** between France and Spain.

____ **4.** The Mid-Atlantic **Ridge** in the Atlantic Ocean is the longest mountain range on Earth.

____ **5.** The ocean floor is not flat. It has many tall ridges and deep **trenches**. For example, in the Pacific Ocean, the Mariana Trench is almost 11,000 meters below the surface of the ocean.

____ **6.** Earth's crust is not solid. It is made up of many different **plates**.

B Match the definitions with the words in **bold** above. Go back to Step A. Write the letter of each definition in the blank.

a. a line that divides two places or areas
b. large pieces of Earth's crust
c. a chain of mountains
d. areas of land completely surrounded by water
e. very large areas of land, often made up of many countries
f. long, narrow, deep holes

Reading 1

PLATE TECTONICS

Earth is always moving. You may not feel it, but our whole planet turns as it orbits the sun. There are movements on Earth's surface, too. Land moves, and mountains grow taller. For example, each year South America moves approximately two centimeters farther away from 5 Africa, the islands of Hawaii move about seven centimeters to the northwest, and Mount Everest slowly rises five millimeters upward. Why are continents, islands, and mountains moving? For many years, scientists did not have an answer.

Look at the seven continents on a map of the world, and you may 10 notice that they seem to fit together like pieces of a puzzle. In 1912, a German scientist named Alfred Wegener thought of an interesting idea. He suggested that millions of years ago, Earth had just one giant continent. He called it Pangaea (pan-GEE-uh). Pangaea means "all the Earth" in Greek. Wegener believed that as time passed, Pangaea broke 15 apart, and the pieces drifted, or moved, to where the continents are today. He called his idea **continental drift theory**, but this idea did not explain how the continents moved. Wegener didn't know what scientists know today. Today, scientists know that the continents move because of **plate tectonics**.

continental drift theory the idea that, over time, the continents move toward or away from each other

Tectonic plates and plate boundaries

In the 1960s, scientists discovered that Earth's crust is broken into large pieces. These pieces are called tectonic plates. No one knows the exact number of plates, but many scientists agree that there are about 12 large plates and several smaller ones. These plates are under the continents (continental plates) and under the oceans (oceanic plates). The plates, and the continents and oceans on top of them, move in different directions and at different speeds. Tectonic plates interact at places called plate boundaries. There are three types of plate boundaries: divergent boundaries, convergent boundaries, and transform boundaries.

plate tectonics the movement of large pieces of Earth's crust and the contact between them

Divergent boundaries

Divergent boundaries are where two plates *diverge*, or move away, from each other. When two oceanic plates diverge, the ocean floor grows wider, and an underwater ridge (mountain range) forms. A good example is the Atlantic Ocean. Millions of years ago, the Atlantic Ocean was a very small body of water. As the plates under it diverged, the ocean grew approximately two centimeters wider each year, and a ridge formed. Today the Atlantic is a huge ocean, and the Mid-Atlantic Ridge is the longest mountain range on Earth.

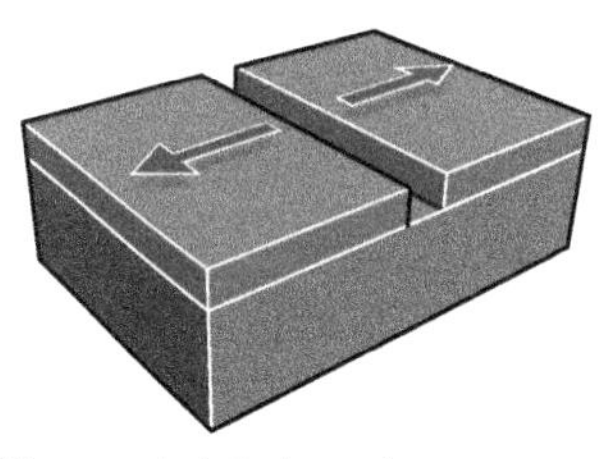

Divergent plate boundary

Convergent boundaries

Convergent boundaries are where two plates *converge*, or come together. When two oceanic plates converge, they form a trench and a group of islands, such as the Mariana Trench and the Mariana Islands in the Pacific Ocean. When an oceanic and a continental plate converge, they create a trench and a mountain range. The Peru-Chile Trench and the Andes Mountains formed in this way. When two continental plates converge, a mountain range forms. This process created the Himalayas, the great mountain range in Asia.

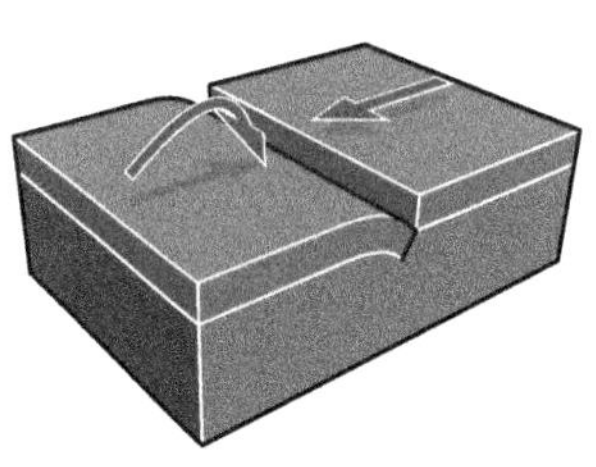

Convergent plate boundary

Transform boundaries

At transform boundaries, two plates slide past each other. As they move, they can bump, or hit, each other. This movement often causes an earthquake, which is a movement of Earth's crust. People who live along the coast of California often experience earthquakes. They are very common at the transform boundary between the Pacific Plate and the North American Plate.

Continental drift continues. Even though the plates move just a few centimeters a year, over a long period of time, they cause Earth to grow and change in dramatic ways.

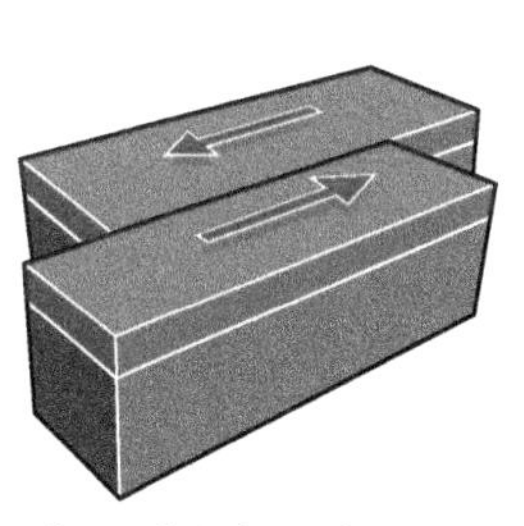

Transform plate boundary

AFTER YOU READ

1 Using headings to remember main ideas 

After you read a text, look back at the headings. The headings will help you remember the main ideas.

A Read these headings from the text "Plate Tectonics."

a. Tectonic plates and plate boundaries
b. Divergent boundaries
c. Convergent boundaries
d. Transform boundaries

B Work with a partner. Match the headings in Step A with the main ideas below. Write the letter of the correct heading in the blank.

____ **1.** Sometimes two plates move away from each other. This often creates ridges.

____ **2.** Sometimes two plates come together, creating mountains, islands, and trenches.

____ **3.** Earth's crust is divided into about 12 large pieces and several smaller ones that move and interact with each other.

____ **4.** When two plates slide by each other, they can bump. This often causes an earthquake.

2 Prefixes V

A **prefix** is a word part that comes at the beginning of a word. Each prefix has a meaning. For example, the prefix *re-* means "again." To *reread* a book means to read it again. To *rewrite* a letter means to write it again. Knowing the meaning of a prefix can often help you guess the meaning of a word.

Prefix	Meaning
con-	together, with
cent-	100
inter-	between two or more things or groups
mil-	1,000

A Work with a partner. Find these words in the text "Plate Tectonics": *centimeters*, *millimeters*, *converge*, *interact*. Circle them. Look at the prefixes and guess the meanings of these words. Use a dictionary if necessary.

B Here are some new words with the prefixes you learned.

century	convention	interplanetary	millennium

Complete the sentences with the correct words from the box.

1. Do you think that someday there will be __________ flights between Earth and Mars?
2. The most powerful earthquake of the past __________ happened in Chile in 1960. It was the strongest earthquake in the last 100 years.
3. This week there is a __________ of astronomers at the university. Hundreds of astronomers are meeting to talk about their research.
4. Many people had parties to celebrate the start of the new __________ in the year 2000.

C Think of more words that start with the prefixes *cent-*, *con-*, *inter-*, or *mil-*. Make a class list.

3 Prepositional phrases (V)

A **prepositional phrase** is a preposition + a noun (or noun phrase) or a pronoun.
Examples: *on Earth's surface, in the 1960s, at different speeds*

Example: **on** (preposition) **Earth's surface** (noun phrase)

Prepositional phrases often answer the questions *Where?*, *When?*, or *How?*

A Find the following prepositional phrases in the text. Underline them. Decide what question each phrase answers. Write *Where?*, *When?*, or *How?* in the blank.

Where? **1.** on Earth's surface

__________ **2.** along the coast of California

__________ **3.** (millions) of years ago

__________ **4.** under the continents

__________ **5.** at different speeds

__________ **6.** in this way

B Work with a partner. Find six more prepositional phrases in the text and underline them. Decide if each phrase answers the question *Where?*, *When?*, *How?*, or none of these.

4 Reading maps (A) (R)

Maps show different places on Earth's surface. They can help you find the places you read about. Most maps have a key that includes information to help you read the map. In addition, the key often has a compass or a drawing that shows the directions: North, South, East, and West.

A Look at the map of the world's tectonic plates on page 28. Work with a partner and find these continents: Eurasia (Europe and Asia), North America, South America, Antarctica, Australia, Africa.

B Read the statements and answer *T* (true) or *F* (false). Use the information from the map.

_____ **1.** The North American Plate is northeast of the Pacific Plate.
_____ **2.** The Pacific Plate is smaller than the African Plate.
_____ **3.** The Nazca Plate is east of the South American Plate.
_____ **4.** The Australian Plate is south of the Indian Plate and the Philippine Plate.
_____ **5.** There is no Atlantic Plate.

C Write three sentences about the map. They may be true or false. Exchange sentences with a partner. Decide if your partner's sentences are true or false.

5 Writing simple and compound sentences W

You have learned that every English sentence must have at least one subject and one verb. A sentence with only one subject and one verb is called a **simple sentence**.

subject: Continental drift | verb: continues | today.

Continental drift continues today.

A sentence with more than one subject, more than one verb, and a coordinating conjunction (connecting word) is called a **compound sentence**. The most common coordinating conjunctions are *and*, *or*, and *but*. There is always a comma before these conjunctions.

subject	verb		coordinating conjunction	subject	verb	
Some plates	are	under the continents,	and	some plates	are	under the oceans.

Try to use both simple and compound sentences in your writing. Vary the sentence structure, and you will make your writing more interesting to read.

A Go back to the text. Find three simple sentences and three compound sentences. Write these sentences on a separate piece of paper. Underline the subjects once and the verbs twice. Circle the coordinating conjunctions. Compare your answers with a partner's.

B Write four or five sentences about what you learned in the reading "Plate Tectonics." Try not to look back at the text. Include simple and compound sentences. You can use these words and phrases: *continental drift theory*, *continents*, *Pangaea*, *tectonic plates*, *divergent boundaries*, *convergent boundaries*, and *transform boundaries*. Compare your sentences in a small group.

PREPARING TO READ

Building background knowledge (R)

> Learning basic facts about the topic of a text builds your knowledge of the topic. This can help tell you what kinds of terms and ideas you will read about.

A Read the following paragraph about volcanoes.

A volcano is a mountain with a hole at the top. When a volcano erupts, it throws smoke, gas, ashes, and lava (melted rock) out of the hole. Some volcanoes, like Mauna Loa in Hawaii, are active. This means that they are erupting or that they could erupt at any time. Other volcanoes are extinct, or dead. Scientists believe these volcanoes will not erupt again.

B Discuss these questions with a partner.

1. Are there any volcanoes where you live? Are they active or extinct?
2. Do you know the names of any famous volcanoes? If so, which ones?
3. Have you ever seen a volcano erupt? If so, where did you see it? What did you see?

C Test your knowledge of volcanoes. Answer these questions with a partner.

1. Which is not a volcano?
   - **a.** Mount Fuji (Japan)
   - **b.** Mount Everest (Nepal)
   - **c.** Mount Vesuvius (Italy)
   - **d.** Krakatau (Indonesia)
2. Which country has no active volcanoes?
   - **a.** Japan
   - **b.** Italy
   - **c.** The United States
   - **d.** Australia
3. Which island was not formed by volcanic activity?
   - **a.** Greenland
   - **b.** Iceland
   - **c.** The Hawaiian Islands (U.S.A.)
   - **d.** Honshu Island (Japan)
4. Which volcano is extinct?
   - **a.** Sangay (Ecuador)
   - **b.** Pinatubo (Philippines)
   - **c.** Kohala (Hawaii, U.S.A.)
   - **d.** Stromboli (Italy)

Reading 2

VOLCANOES

One afternoon in 1943, a farmer in Paricutín, Mexico, went to his cornfield. In the cornfield, he saw something unusual. It was a hole in the ground with smoke coming out of it. The next day, there was a 10-meter hill in the same place. Rocks were flying from the hilltop, 5 and lava was flowing down its sides. After one year, the hill was 450 meters high, and it continued to erupt. The farmer was amazed, and frightened, too. He was watching the birth of a volcano.

The formation of volcanoes

When the magma under Earth's crust breaks through to the surface, it creates a volcano. Volcanoes usually form at plate boundaries, where 10 the crust is the weakest. More than 75 percent of Earth's volcanoes are located around the Pacific Plate, in a region called the Ring of Fire. In the Atlantic Ocean, there are many volcanoes at the boundary between the North American Plate and the Eurasian Plate. Directly on top of the two diverging plates is the volcanic island of Iceland.

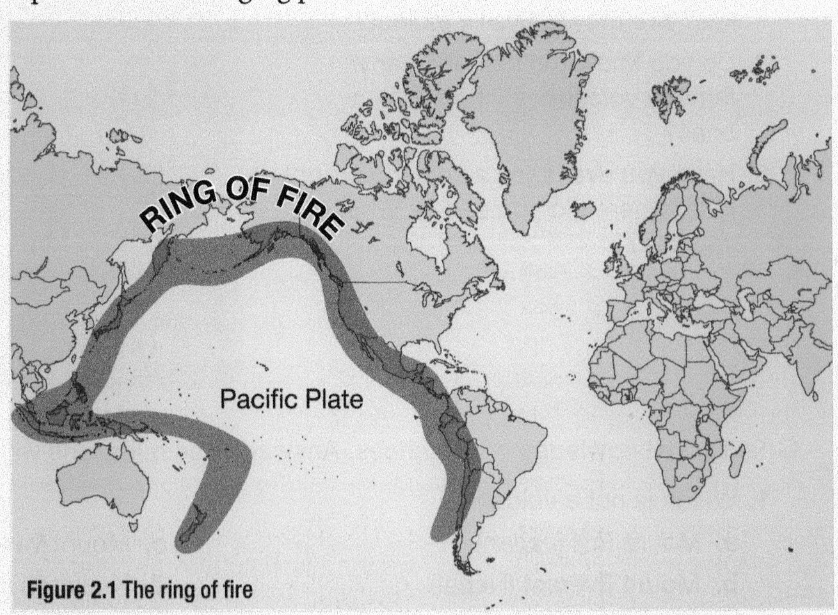

Figure 2.1 The ring of fire

15 Most volcanoes form near plate boundaries, but a few do not. Instead, they form in the middle of a plate over a hot spot (a stream of hot magma that is deep inside Earth). This magma flows up and breaks through the tectonic plate above it. The lava from the hot spot eventually creates a volcanic island. Hot spots do not move, but 20 tectonic plates do. When the plate moves over the hot spot, the volcano also moves, so it stops erupting. Nearby, a new volcanic island forms over the hot spot. Over millions of years, this process results in a whole chain of islands. This is how the Hawaiian Islands were formed.

Active and extinct volcanoes

In the world today, there are approximately 1,500 active volcanoes. 25 Active volcanoes are volcanoes that are erupting or that might erupt in the future. Active volcanoes can be extremely destructive. Lava, gases, **ash**, and rocks can suddenly erupt from a volcano and destroy everything around it. One active volcano is Mount Vesuvius. Mount Vesuvius erupted in 79 CE and buried the city of Pompeii, Italy, in ash. 30 Mount Tambora (Indonesia, 1815), Mount Krakatau (Indonesia, 1883), and Mount Pelée (Martinique, 1902) also caused major destruction.

ash the soft gray or black powder that is left when something burns

Some volcanoes will not erupt again. These are called *extinct volcanoes*. Scientists generally agree that Hawaii's oldest volcano, Kohala, is extinct.

35 Volcanoes can be destructive and deadly. However, they can also have a positive effect on Earth. Volcanoes form new mountains, new islands, and new land. In this way, volcanic activity is an important natural process that contributes to our planet's growth.

The Year Without a Summer

Many scientists believe that the eruption of Mount Tambora in Indonesia in 1815 was the most destructive volcanic eruption in the past 10,000 years. People heard the explosion 2,600 kilometers away. One hundred fifty cubic kilometers of ash fell within the first 24 hours of the eruption. Huge amounts of ash filled the air. Over the next year, the large amount of ash in the air blocked the sun's rays from reaching Earth. This had a dramatic effect on weather patterns on the other side of the world.

In New England and Canada, snow fell during the summer of 1816. Cold weather killed farmers' crops and caused serious food shortages. In Europe, the situation was much worse. Cold weather and heavy rains caused famine, or great hunger, in France and Switzerland. In Ireland, cold rain fell almost every day that summer. Thousands of people were hungry, and many got a disease called typhus. This terrible illness spread to other parts of Europe, and all of this trouble started with the volcanic eruption.

AFTER YOU READ

1 Answering true / false questions (A)

True / false questions are common on tests and in textbooks. Here are some guidelines for answering this type of question:

- Most activities and tests with true / false questions have approximately the same number of true sentences and false sentences.
- Sentences with words like *never*, *always*, *only*, and *all* are often false.
- Sentences with words like *often*, *many*, and *sometimes* are often true.

A Write *T* (true) or *F* (false) for each statement. Find the answers in the reading "Volcanoes."

_____ **1.** Volcanoes often form at places where Earth's crust is weak. Par. _____

_____ **2.** The Ring of Fire is in the Atlantic Ocean. Par. _____

_____ **3.** Iceland is directly on top of two divergent plates. Par. _____

_____ **4.** Hot spots are not located near plate boundaries. Par. _____

_____ **5.** One giant volcanic eruption formed the Hawaiian Islands. Par. _____

_____ **6.** Today the world has approximately 15,000 active volcanoes. Par. _____

_____ **7.** Mount Vesuvius in Italy is an active volcano. Par. _____

_____ **8.** Volcanoes always have a negative effect on Earth. Par. _____

B Now go back to Step A. Write the number of the paragraph where you found each statement.

C Work with a partner. Find the sentences you marked false in Step A and correct them.

2 Writing definitions (W)

Writing the definition of a word helps you remember the word. Here are two common ways to write a definition:

1. (A) __________ is / are (a) __________ .
A plate **is a** large piece of Earth's crust.
Marble **is a** metamorphic rock.
Plates **are** large pieces of Earth's crust.

2. (A) __________ is / are a __________ that . . .
A planet **is a** large object **that** orbits the sun.
Planets **are** large objects **that** orbit the sun.

A Complete the definitions of the words in **bold** below. Use the definitions in Reading 2 "Volcanoes," but try not to look back at the text.

1. A **hot spot** is a ______________________________.

2. An **active volcano** is a volcano that ______________________________.

B Complete the definitions of the words in **bold** below. Use the definitions in Reading 1 "Plate Tectonics," but try not to look back at the text.

1. Tectonic plates are ______________________________.

2. A **ridge** is ______________________________.

3. An **earthquake** is a ______________________________.

C Compare your definitions in Steps A and B with a partner. Now look back at the texts to check your work.

3 Reading boxed texts Ⓡ

Many academic textbooks include boxed texts. Boxed texts usually contain interesting material that will add to your understanding of the main text.

Boxed texts can have different purposes. For example, they may:

- give interesting examples of ideas in the main text
- give more details about a topic in the main reading
- discuss a topic that is closely related to the topic of the main text
- help you apply the information from a text to your own life
- present a point of view, or way of thinking about something, that is different from the one in the main text

A Reread the boxed text "The Year Without a Summer" on page 35.

B In a small group, discuss the purpose of the boxed text. Does it match any of the purposes in the box above?

C Go back to the boxed texts in Chapter 1 on pages 6 and 20. What is the purpose of each boxed text? Write your answers below.

- "The Story of Pluto": ______________________________.
- "Save the Rocks!": ______________________________.

4 Illustrating main ideas R A

Drawing a picture of a key point from the text can be a useful way to take notes and check your understanding of the reading.

A Reread paragraph 1 of "Volcanoes."

B Look at the illustration below. This shows what happened the first day the farmer went into his cornfield.

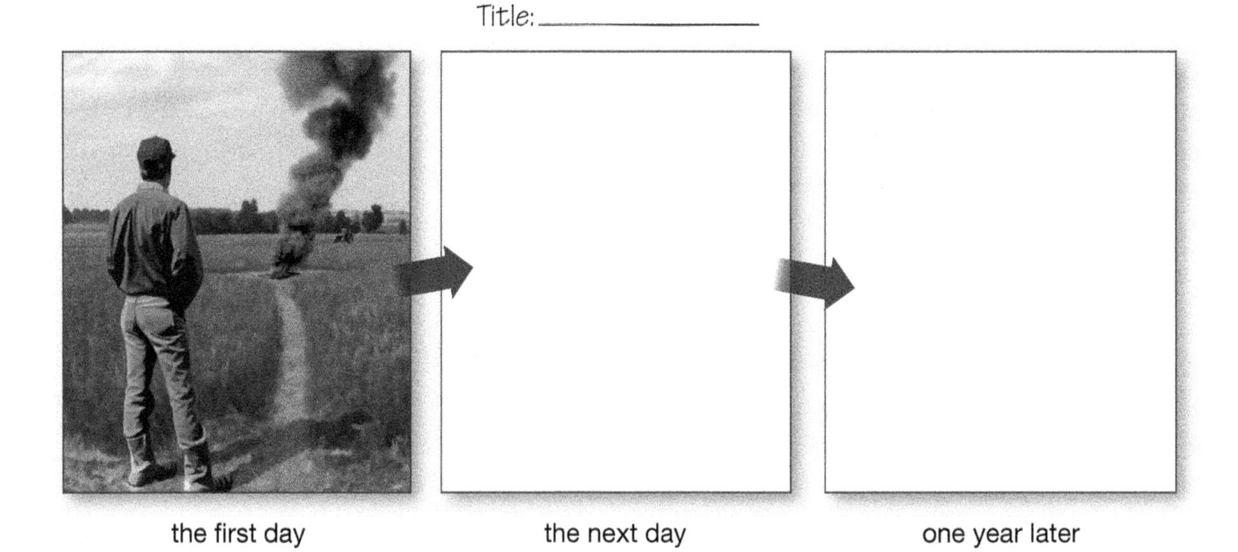

the first day | the next day | one year later

C In the second box, draw a picture of what happened the next day. In the third box, draw what happened one year later. When you are finished, give your set of illustrations a title.

D Choose another section of the text to illustrate. Use a separate piece of paper to make your illustrations.

E Share your illustrations in a small group. Guess which idea each illustration represents. Do they help you understand the ideas in the text better? Which illustration do you like the best? Why?

PREPARING TO READ

Thinking about the topic Ⓡ

A Work with a partner. Look at the photographs and answer the questions below.

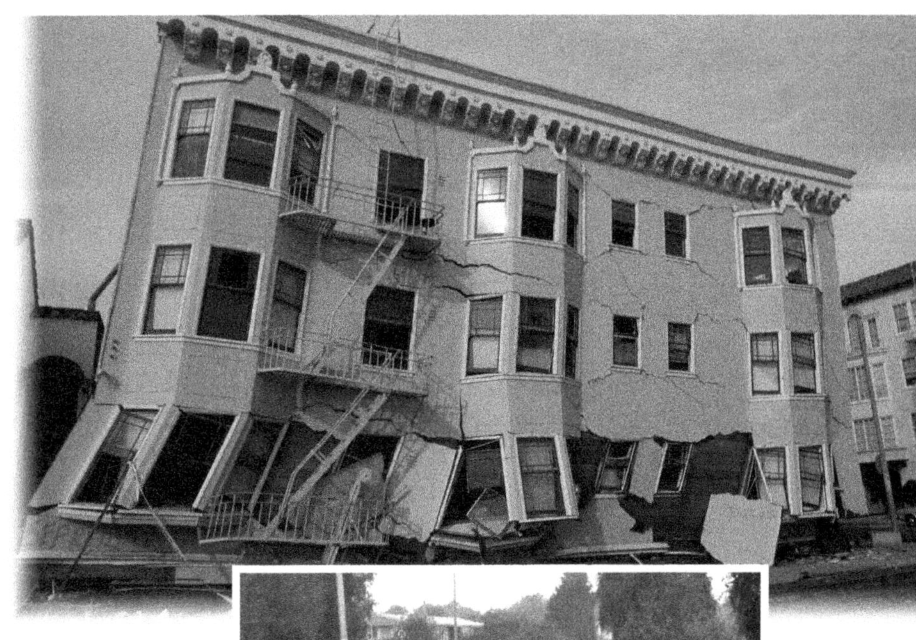

1. What do you see in the photographs?
2. What do you think happened?
3. Where and when do you think the photographs were taken?

B Discuss the following questions in a small group.

1. What is an earthquake?
2. Why do some places, such as California and Japan, have so many earthquakes?
3. Does the place where you live have earthquakes?
4. Have you ever been in an earthquake? If so, describe your experience.
5. What are three things people can do to stay safe during an earthquake?

Reading 3

EARTHQUAKES

As 1974 came to an end and the new year began, animals in Haicheng, China, started acting strangely. Snakes normally hibernate underground during the winter, but they suddenly came out of their holes. Dogs began to bark and run around wildly, and horses became 5 so upset that some ran away. Why were the animals acting like this? Many people think that the animals sensed what was coming: On February 4, 1975, the earth began to shake and buildings collapsed as a very large earthquake struck the city of Haicheng.

What causes earthquakes?

When the tectonic plates that make up Earth's crust move past each 10 other, they often bump or rub against each other. The earth, or ground, above the plates moves as well. This movement is called an earthquake. Sometimes the plates get stuck. The pressure increases as the two plates try to move past each other but cannot. They finally move with a sudden and powerful jerk. This can also cause an 15 earthquake. During a small earthquake, the ground simply shakes a little, and people may not even notice. However, a strong movement can cause the ground to shake and roll violently. It can make buildings and bridges fall. It can also cause the ground to split open and form a large **fault**, or crack.

fault a large break in the surface of the earth

Where do earthquakes happen?

20 Earthquakes can happen anywhere, but certain places have more earthquakes. These places sit on tectonic plates that move frequently. One example is the area around the Pacific Plate, which includes China, the Philippines, Japan, and the western coasts of Canada, the United States, and South America. Earthquakes are common in 25 those places. The deadliest earthquake in modern times happened in 1976 in Tangshan, China. It lasted less than two minutes, but more than 250,000 people died, and more than 90 percent of the buildings collapsed. Earthquake scientists study places such as Tangshan because of the many faults in these areas and the activity of the tectonic plates.

San Andreas fault in California

Can people prepare for an earthquake?

30 Scientists cannot predict when an earthquake will happen. However, they are able to tell us the areas where earthquakes are most likely to happen. This information helps people in those areas to prepare. They can learn what to do before, during, and after an earthquake. Engineers in those areas can build bridges and buildings that are 35 better able to survive earthquakes. We cannot stop tectonic plates from moving, but with accurate information and good planning, we can help people live more safely on our planet.

Still Standing: Earthquake-Resistant Buildings

Humans can be very daring. They swim with sharks and run with bulls. They fly to the moon and dive in the ocean. They also build very tall buildings. In 2010, the Burj Khalifa tower (828 meters) in the United Arab Emirates took the title of world's tallest building away from Taipei 101 (508 meters) in Taiwan. Most likely, someone will soon build a tower taller than Burj Khalifa. However, as people build taller and taller buildings, they must also consider how an earthquake might affect these enormous skyscrapers.

Earthquakes can be very destructive. Therefore, people who design buildings try to find ways to make even the tallest buildings stronger and more flexible. They hope their new designs will make the buildings more earthquake resistant. For example, Taipei 101 has a giant steel ball hanging in the middle of the building. If an earthquake causes the building to sway, the ball will move in the opposite direction. This keeps Taipei 101 balanced and helps stop it from falling down.

Another good example of an earthquake-resistant building is the Sabiha Gökçen International Airport in Turkey. The main building of the airport does not touch the ground. Instead, it sits on top of 300 rubber and steel bearings. When an earthquake shakes the ground, the balls let the tower move side to side and back and forth. Without the rollers, the walls of the building would shake, and it could collapse.

Inside Taipei 101

AFTER YOU READ

1 Reading for main ideas (R)

It is very important to learn how to find the main idea of a paragraph or longer text. To identify the main idea of a text, ask "What is it about?" or "What idea do all the sentences discuss?"

A Read the sentences. These are main ideas in "Earthquakes." Find each idea in the reading and then write the number of the paragraph that discusses it.

1. Some places experience more earthquakes than others. Par. ____
2. Animals may have predicted the earthquake that struck Haicheng, China, in February of 1975. Par. ____
3. Earthquakes happen when two tectonic plates bump or get stuck as they move past each other. Par. ____
4. Scientists cannot predict when an earthquake will happen, but their information can help people prepare for one. Par. ____

B Check (✓) the sentence that expresses the main idea of the whole text.

____ **1.** A serious earthquake occurred in Haicheng, China, in 1975.

____ **2.** Earthquakes, caused by the movement of tectonic plates, can happen anywhere, and people need to prepare for them.

____ **3.** It is impossible to prepare for earthquakes because no one knows when they will happen.

____ **4.** Earthquakes usually last a very short period of time, but they can kill thousands of people and cause buildings to collapse.

2 Using grammar, context, and background knowledge to guess meaning V

Sometimes you can use grammar, context, and background knowledge to guess the meaning of new words.

Grammar: Look at the part of speech. It tells you if the new word is a thing (noun), an action (verb), or a descriptive word (adjective).

Context: Look at the words and sentences before and after a new word. They often include a definition or a description that can help you guess the meaning.

Background knowledge: Consider what you might already know about a new word. You may recognize a word part or know something about the context's topic.

A Read the following paragraph from "Earthquakes."

As 1974 came to an end and the new year began, animals in Haicheng, China, started acting strangely. Snakes normally **hibernate** underground during the winter, but they suddenly came out of their holes. Dogs began to **bark** and run around wildly, and horses became so **upset** that some ran away. Why were the animals acting like this? Many people think that the animals **sensed** what was coming: On February 4, 1975, the earth began to shake and buildings **collapsed** as a very large earthquake struck the city of Haicheng.

B Work with a partner. Match the words in **bold** in Step A with the definitions below. Use the strategies for guessing meaning.

Example: Snakes normally **hibernate** underground during the winter.

Strategies

- grammar — You can guess that the word *hibernate* is a verb (an action).
- context — The other words in the sentence tell you that hibernate is something that animals do underground during the winter.
- background knowledge — You may know that some animals sleep during the winter.

hibernate **a.** sleep during the winter
_________ **b.** felt something without seeing or hearing it
_________ **c.** make a loud animal noise
_________ **d.** fell down
_________ **e.** worried, unhappy

C Discuss your answers in a small group. Tell which strategies you used to guess each word.

3 Pronoun reference W R

Pronouns are words that take the place of nouns. Read the sentence below. Notice that the pronoun *he* refers to "professor." The pronoun *it* refers to "test."

Next week, my professor will give us a test on plate tectonics. He told us that we should review our notes to prepare for it.

Skillful writers use pronouns to replace nouns. Pronouns can add interest and help connect sentences and ideas. Find the noun that a pronoun refers to, and you will understand the meaning of the pronoun.

Follow these rules to find the noun that a pronoun refers to:

- Look for the noun that comes before the pronoun. Pronouns usually come after the nouns that they refer to.
- Notice if the pronoun is singular or plural. That will tell you to look for a singular noun or a plural noun.

A Read these sentences from the text "Earthquakes." The pronouns are underlined. Draw an arrow from each pronoun to the noun or noun phrase it refers to.

1. When the tectonic plates that make up Earth's crust move past each other, they often bump or rub against each other.
2. The pressure increases as the two plates try to move past each other but cannot. They finally move with a sudden and powerful jerk.
3. However, a strong movement can cause the earth to shake and roll violently. It can make buildings and bridges fall. It can also cause the earth to split open and form a large fault, or crack.
4. The deadliest earthquake in modern times happened in 1976 in Tangshan, China. It lasted less than two minutes, but more than 250,000 people died, and more than 90 percent of the buildings collapsed.

B Compare your work with a partner's.

4 Showing contrast (W)

Writers contrast (show the difference between) ideas with words such as *however* and *but*.

The word *however* often starts a sentence. There is always a comma after it.

An active volcano may not have erupted in thousands of years. However, it could erupt sometime in the future.

The word *but* is a coordinating connector. It comes in the middle of a sentence and has a comma before it. It contrasts two ideas in a compound sentence.

Most volcanoes form near plate boundaries, but a few do not.

A Complete these sentences with *but* or *however*. Then find the sentences in the text "Earthquakes" and check your answers.

1. During a small earthquake, the earth simply shakes a little, and people may not even notice. ________, a strong movement can cause the earth to shake and roll violently.
2. Earthquakes can happen anywhere, ________ certain places have more earthquakes because they sit on tectonic plates that move frequently.
3. Scientists cannot predict when an earthquake will happen. ________ , they are able to identify the areas where earthquakes are most likely to occur.
4. We cannot stop tectonic plates from moving, ________ with accurate information and good planning, we can help people live more safely on our planet.

B Write your own sentences with *but* and *however*. Follow the example. Use correct punctuation and capital letters where necessary. Use a separate piece of paper.

1. Volcanoes are destructive.

   Volcanoes are destructive, but they also create new land on Earth.

   Volcanoes are destructive. However, they also create new land on Earth.
2. California has earthquakes every day.
3. Small earthquakes do not shake the ground very much.
4. You may not feel the ground move.

Chapter 2 Academic Vocabulary Review

The following words appear in the readings in Chapter 2. They all come from the Academic Word List, a list of words that researchers have discovered occur frequently in many different types of academic texts. For a complete list of all the Academic Word List words in this chapter and in all the readings in this book, see the Appendix on page 206.

accurate	dramatic	major	region
collapsed	eventually	normally	survive
create	interact	predict	theory

Complete the sentences with words from the list.

1. One ________ of the world that has a lot of earthquakes is the west coast of South America.
2. Volcanic eruptions can ________ many problems for the people who live nearby. The lava can destroy houses, and the ash can cause breathing difficulties.
3. ________ people do not feel the earth move beneath their feet, but sometimes they do feel an earthquake shake the ground.
4. Did huge volcanic eruptions kill the dinosaurs on our planet? That is one idea. Another ________ is that the dinosaurs died because an asteroid hit Earth.
5. It is important not to panic in an earthquake. You can ________ an earthquake, but you must stay calm and get to a safe place.
6. When scientists collect data, it is important that the information is ________ . Incorrect information could ruin their research.
7. Scientists study the recent activity of a volcano to try to ________ when it will erupt.
8. The roof of the house ________ when a large rock fell on it.
9. It is often useful to ________ with your classmates outside of school. You can make new friends and help each other study.
10. There is one ________ problem with staying up all night to study for a test. You are often too tired to remember the information the next day!

Practicing Academic Writing

In Unit 1, you learned about planet Earth and its physical features. Based on everything you have read and discussed in class, you will write a paragraph about this topic.

A Special Place

You will write one academic paragraph about a place on Earth that you like. You might choose a place that you think is beautiful, interesting, or fun.

PREPARING TO WRITE

1 Using correct paragraph form

A **paragraph** is a group of sentences about the same topic. In general, paragraphs in English have six to ten sentences, but they can be shorter or longer. When you write a paragraph, follow these rules of form:

- Indent the first sentence of a paragraph. That means, start the first sentence five spaces from the left margin. Sometimes you will see paragraphs that do not follow this rule. However, in your writing, you should always indent the first sentence.
- Begin each sentence with a capital letter. End each sentence with punctuation such as a period or question mark.
- Write one sentence directly after another sentence. Do this until you get to the end of a line. Do not use a separate line for each sentence.

A Look at the following paragraph. With a partner, identify the three ways this paragraph uses correct form.

Look at the seven continents on a map of the world, and you may notice that they seem to fit together like pieces of a puzzle. In 1912, the German scientist Alfred Wegener suggested that millions of years ago, Earth had just one giant continent. He called it *Pangaea*, which means "all the Earth" in Greek.

B The following text does not follow the rules of correct paragraph form. Rewrite the text as a paragraph. Use a separate piece of paper. Use correct paragraph form and correct punctuation.

There are four basic types of volcanoes: shield volcanoes,
composite volcanoes, cinder cone volcanoes, and supervolcanoes.
shield volcanoes are generally very large, and lava usually flows down their sides.
Composite volcanoes are smaller than shield volcanoes.
They can have both small eruptions and big eruptions
The smallest type of volcano is the cinder cone volcano.
For example, the Paricutín volcano was a cinder cone volcano.
the largest and most dangerous volcanoes are supervolcanoes, and they can
cause a lot of destruction.
Scientists continue to study these four types of volcanoes to learn more about our planet.

C Compare your paragraph with a partner's.

2 Using correct paragraph structure

In addition to following correct paragraph form, academic paragraphs often have a specific structure:

- The first sentence of the paragraph is the **topic sentence**. It explains the main idea of the whole paragraph.
- The middle sentences of a paragraph are **supporting sentences**. They give details and examples that explain the main idea.
- The final sentence, or **concluding sentence**, ends the paragraph by reminding the reader about the main idea of the paragraph.

Not all academic paragraphs follow this structure, but many do. Be sure to learn this pattern. It will help you read and understand academic texts.

A Reread paragraph 3 of the text "Earthquakes." Underline and label the topic sentence (*TS*). Next, bracket ([ ]) and label the supporting sentences (*SS*). Then draw two lines under the concluding sentence and label it (*CS*).

B Read the following paragraph.

There is no way to stop an earthquake, but there are several things you can do to prepare and protect yourself. Before an earthquake happens, you should make an emergency plan. You should also prepare an emergency supply kit with a battery-powered radio, a flashlight, and enough food and water for three days. Remember to do these things during an earthquake: Stay away from windows and tall furniture inside a building. Get on the floor, cover your head, and hold on to something until the shaking stops. Find a place away from buildings and trees outside and get on the ground. After the earthquake stops, check for injuries – are you hurt? Listen to the radio for instructions. If you are in an unsafe building, go outside. An earthquake can be a frightening experience, but knowing what to do before, during, and after it will help you stay safe.

C Work with a partner. Answer these questions about the paragraph in Step B.

1. Does the paragraph have a topic sentence? If so, underline it, and label it (*TS*).
2. How many supporting sentences are there? Bracket ([ ]) and label them (*SS*).
3. Does the paragraph have a concluding sentence? If so, draw two lines under it, and label it (*CS*).

NOW WRITE

Writing first drafts

Your first piece of writing on a topic is a first draft. Very few people write a "perfect" first draft. A first draft gets some ideas and sentences down on paper so you can then read the paragraph and find ways to improve it.

A Think about your favorite place on Earth. It can be a continent, an island, a mountain range, a river, a lake, or some other place. Consider places that you visited or learned about in a book, a movie, or on the Internet.

B Write about this place on a piece of paper for five minutes. Do not stop to erase ideas or correct mistakes. Just freewrite about the place without stopping. The purpose of this task is to write down as many thoughts and ideas as you can.

C Have a short conversation with a partner. Tell each other about your favorite places.

D Now write a paragraph about your favorite place.

1. Use this topic sentence to begin. Choose the word that best describes your place.

   topic sentence: *The most (interesting / exciting / beautiful) place on Earth that I know is* ________________ .

2. Write five or six **supporting sentences** that show why your topic sentence is true.
3. Complete this sentence to end your paragraph.

   concluding sentence: *For all these reasons,* ________________ *is my favorite place.*

4. Remember to use correct paragraph form and structure.
5. Try to include some of the new vocabulary that you learned in this chapter.
6. Try to vary your sentence structure. Use simple and compound sentences. Try to include prepositional phrases, and use pronoun references correctly.

AFTER YOU WRITE

After you complete your first draft, you can revise it. When you revise, you find ways to make your writing better. Often, it helps to have another person read your writing and offer suggestions for improvement.

A Exchange paragraphs with a partner and read each other's work. Then discuss the following questions.

1. Are your favorite places similar or very different?
2. Does your partner's paragraph use correct paragraph form? Are there any problems with form that need to be corrected?
3. Does your partner's paragraph have a clear topic sentence at the beginning and a clear concluding sentence at the end? If not, help your partner write one.
4. Do the supporting sentences explain the topic sentence? Are there any sentences that do not? Cross off any sentences that do not support the topic sentence. Do you think the paragraph has enough supporting sentences? If not, help your partner write more sentences.

B Make any changes to your paragraph that you think will improve it.

C Rewrite your paragraph.

Unit 2

Water on Earth

In this unit, you will look at the importance of water to life on Earth. In Chapter 3, you will examine the natural process that continues to recycle water on our planet. You will also focus on the sources of freshwater, such as rivers, lakes, and glaciers. In Chapter 4, you will discuss Earth's oceans and the activity of currents and waves.

Contents

In Unit 2, you will read and write about the following topics.

Chapter 3 Earth's Water Supply	Chapter 4 Earth's Oceans
Reading 1 The Water Cycle	**Reading 1** Oceans
Reading 2 Groundwater and Surface Water	**Reading 2** Currents
READING 3 Glaciers	**Reading 3** Waves and Tsunamis

Skills

In Unit 2, you will practice the following skills.

R Reading Skills

Thinking about the topic
Examining graphics
Sequencing
Reading about statistics
Increasing reading speed
Reading for main ideas
Scanning
Building background knowledge about the topic
Reading maps
Brainstorming
Reading for main ideas and details

W Writing Skills

Identifying topic sentences
Identifying topic sentences and supporting sentences
Writing topic sentences and supporting sentences
Writing about superlatives
Describing results
Concluding sentences
Parallel structure
Both . . . and and *neither . . . nor*
Reviewing paragraph structure

V Vocabulary Skills

Antonyms
Suffixes that change verbs into nouns
Countable and uncountable nouns
Subject-verb agreement
Too and *very*
Adjective suffixes

A Academic Success Skills

Understanding test questions
Answering multiple-choice questions
Mapping
Conducting a survey
Taking notes
Labeling a map
Organizing ideas

Learning Outcomes

Write an academic paragraph about a water feature on Earth

Previewing the Unit

Previewing means looking at one thing before another. It is a good idea to preview your reading assignments. Read the contents page of every new unit. Think about the topics of the chapters. You will get a general idea of how the unit is organized and what it is going to be about.

Read the contents page for Unit 2 on page 52 and do the following activities.

Chapter 3: Earth's Water Supply

A Look at the photographs and answer the question.

a. ____________________ **b.** ____________________ **c.** ____________________

The photographs show types of freshwater features. Do you know their names? Write the names that you know under the appropriate photographs.

B Discuss these questions with a partner.

1. How are the water features above similar and different?
2. What are some important water features where you live?
3. Have you visited a famous water feature? Talk about your experience.

Chapter 4: Earth's Oceans

A Discuss the following questions in a small group.

1. Is swimming in an ocean different from swimming in other places?
2. Did you ever try sailing, surfing, or scuba diving?
   Describe the ocean at that time. How did it look, smell, and sound?

B The words on the left are connected to oceans. Match the words and definitions.

____ **1.** the Pacific	**a.** winds that blow from west to east
____ **2.** salty	**b.** a famous explorer who named the Pacific Ocean
____ **3.** tsunami	**c.** the taste of ocean water
____ **4.** Magellan	**d.** Earth's largest ocean
____ **5.** the westerlies	**e.** a very big and dangerous ocean wave

Chapter 3

Earth's Water Supply

PREPARING TO READ

1 Thinking about the topic Ⓡ

A Read this information from the text "The Water Cycle." Then answer the questions with a partner.

> We call our planet *Earth*, but many people say that we should call it *Water*. Water covers more than 70 percent of our planet. Water is essential to life on Earth. We drink it, swim in it, clean with it, and use it in many other ways. Surprisingly, the amount of water on Earth does not decrease even though we use so much of it every day.

1. Is *Water* a good name for our planet? Why or why not?
2. Where is water on our planet? Name some places.
3. Use the context to guess the meaning of these words: *essential* and *decrease*.

B People use a lot of water every day. How much water do you use?
Make a list and compare your list with your partner's. Discuss these questions:

1. Who uses more water?
2. Are you surprised at how much water you use in one day?
3. Why do you think the amount of water on our planet never decreases?

2 Examining graphics Ⓡ

A You are going to read about Earth's water cycle. First, look at the diagram of the water cycle on page 56. Then think of what you learned in Chapter 1 about the rock cycle. What does *cycle* mean?

B With a partner, study the diagram of the water cycle again. What is the water cycle? Try to complete the following sentence.

The water cycle is . . .

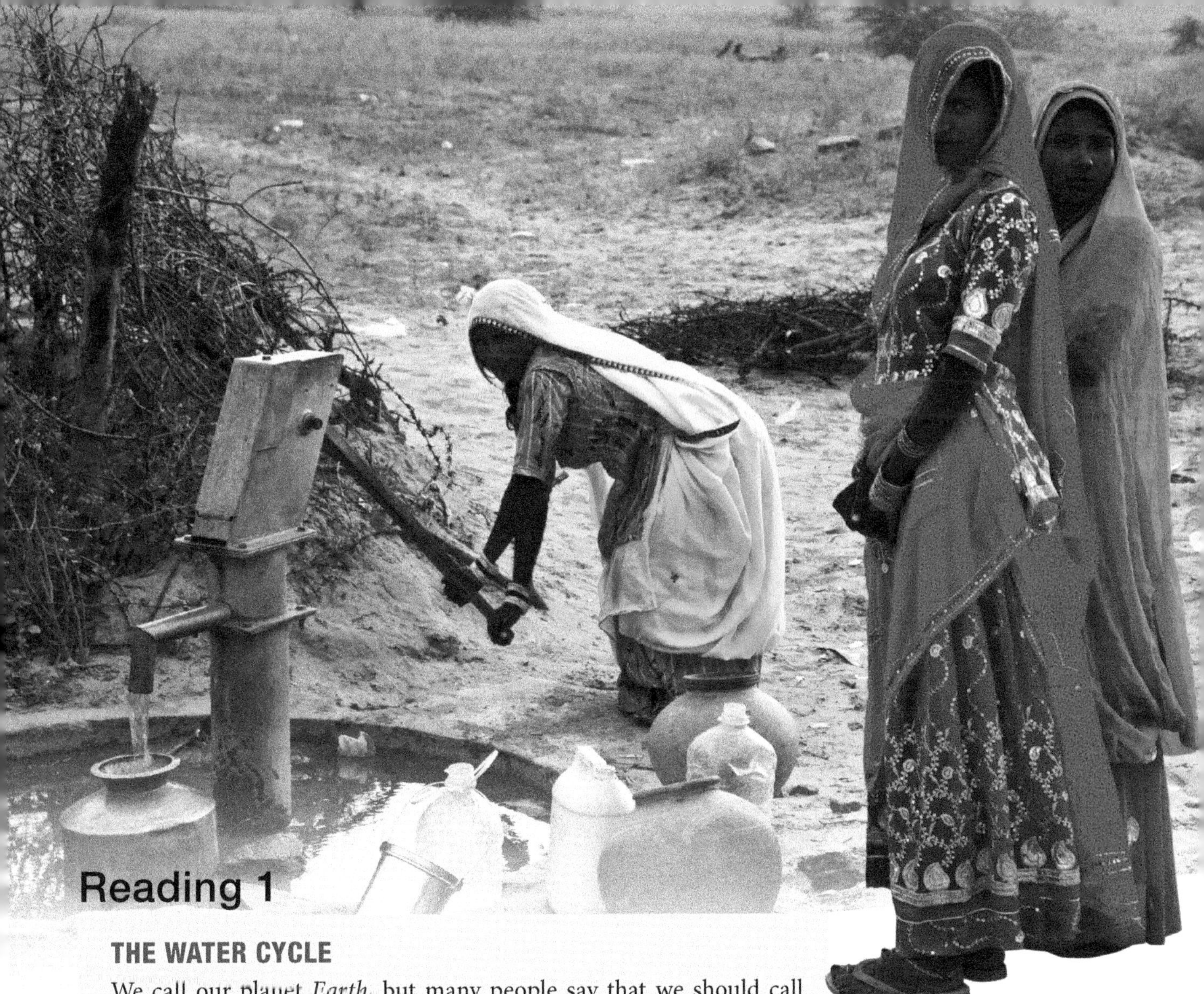

Reading 1

THE WATER CYCLE

We call our planet *Earth*, but many people say that we should call it *Water*. Water covers more than 70 percent of our planet. Water is essential to life on Earth. We drink it, swim in it, clean with it, and use it in many other ways. Surprisingly, the amount of water on Earth does not decrease even though we use so much of it every day. This is because nature recycles water in a process called *the water cycle* (also called *the hydrologic cycle*). The water cycle is the movement of water from Earth into the atmosphere and back to Earth again.

What are the steps of the water cycle?

Evaporation is the first step in the water cycle. This is the process that changes water from a liquid to a gas. Energy from the sun produces **evaporation**. When the sun heats water, some of the water turns into a gas called *water vapor*. Water evaporates anywhere there is sun and water. Most evaporation of water on Earth is from the oceans, but there is also evaporation from lakes, rivers, and even from wet skin and clothing.

evaporation the process that changes a heated liquid to a gas

condensation the process that changes water vapor into droplets of water when it cools

precipitation rain, snow, or hail

Condensation is the second step in the water cycle. Water vapor rises into the atmosphere. It cools and changes back into droplets (very small drops) of liquid water. This process is called **condensation**. When water vapor condenses, it forms clouds.

20 The third step in the water cycle is **precipitation**. Water droplets combine (join together) to form larger drops. The larger drops fall to earth as rain, snow, or hail. Some of this water goes into the ground, and some of it goes into lakes, rivers, and oceans. Eventually, the water that returns to Earth will evaporate and rise into the atmosphere, and 25 the water cycle will continue.

How long is the water cycle?

The fastest water cycle on Earth occurs in tropical rain forests that are near the equator. Tropical rain forests are wet environments. The whole water cycle happens in just one day. In contrast, the slowest water cycle occurs in deserts. Deserts are very dry. It may not rain for 30 years in a desert, so it can take years to go through the whole cycle.

Water on Earth is always moving. It flows down rivers, travels across oceans, evaporates into the atmosphere, and falls to Earth as rain, snow, or hail. The total amount of water on Earth stays the same year after year because of the water cycle. In fact, the water on Earth 35 now is the same water that was on our planet millions of years ago. This means that the glass of water you drink today is millions of years old.

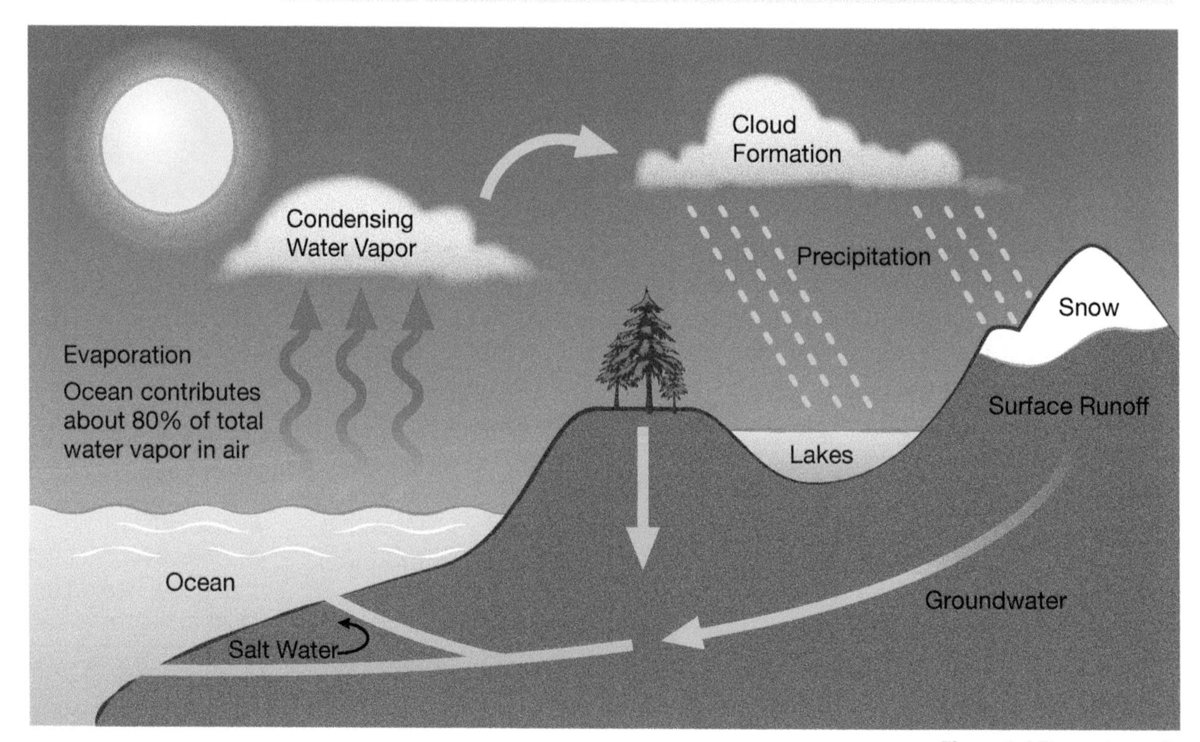

Figure 3.1 The water cycle

AFTER YOU READ

1 Understanding test questions A R

College courses often test you on what you read in the course. It is important to think carefully about each question on the test. First, identify the key words, or important words, in the question. Then decide what the question is asking. Here are some guidelines to help you:

Where questions ask for names and descriptions of places.

Q: Where is the Amazon River?

A: It's in South America.

Q: Where does the slowest water cycle occur?

A: It occurs in deserts, which are very dry.

Why questions ask for explanations or reasons.

Q: Why does the amount of water on Earth always stay the same?

A: The amount of water stays the same because nature recycles it.

How many or *how much* questions usually ask for amounts or numbers.

Q: How much of our planet is covered by land?

A: Less than 30 percent of our planet is covered by land.

A Read each question. Underline the key words. What question is being asked? Use the strategies in the box to decide. Then find the answer in the reading.

1. How much of Earth is covered by water?
2. How many steps does the water cycle have?
3. Where does water evaporate from?
4. Why does water vapor change back to liquid water?
5. Where is the fastest water cycle on Earth?
6. Why is the water on Earth today actually millions of years old?

B Now write two more questions about the text on a separate piece of paper.

C Exchange your questions with a partner and answer each other's questions.

2 Sequencing R

> Scientific information often includes a process, or a series of steps. To understand a process, you need to understand each of the steps and the correct sequence, or order, of the steps.

A Reread paragraphs 2–4 of "The Water Cycle" and review Figure 3.1. Then work with a partner. Look at the steps of the water cycle listed below. Put the steps in order (1–8). Try not to look back at the text.

____ **a.** The water vapor moves up into the atmosphere.

____ **b.** Some of the raindrops fall into lakes, rivers, and oceans.

____ **c.** The sun comes out and begins to warm the water in the ocean again.

____ **d.** The water vapor cools and changes into droplets of water.

____ **e.** The small water droplets inside the cloud combine into bigger water drops, and the bigger water drops fall from the cloud as rain.

____ **f.** Some of the water in the ocean becomes water vapor.

__1__ **g.** The sun heats the water in an ocean.

____ **h.** A cloud forms.

B Complete the diagram of the water cycle. Draw the sequence in Step A and label the steps. Include arrows to show the process.

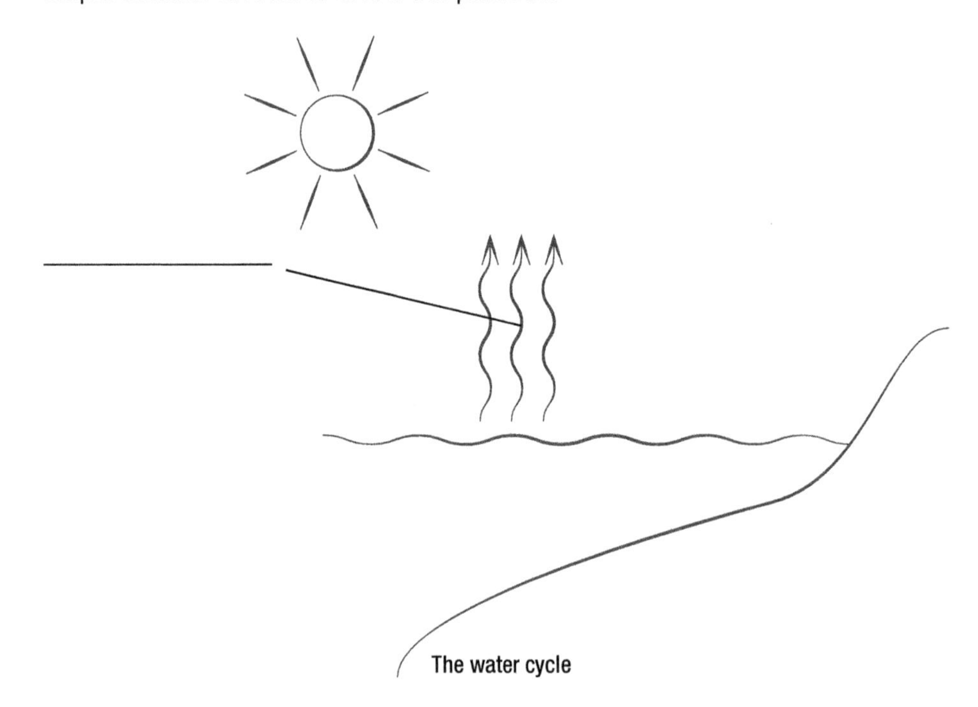

The water cycle

C Look back at Figure 3.1 on page 56 to check your work.

3 Antonyms V

Antonyms are words that have opposite meanings. For example, *cold* and *hot* are antonyms. Learning antonyms is a good way to add words to your vocabulary.

A All of the words below are from the reading. Match the antonyms.

____	**1.** cools	**a.** slowest
____	**2.** rise	**b.** large
____	**3.** evaporation	**c.** dry
____	**4.** fastest	**d.** heats
____	**5.** wet	**e.** fall
____	**6.** small	**f.** condensation

B Choose three pairs of antonyms from Step A. On a separate piece of paper, use the antonyms to write sentences with *but* and *however*. Follow this example:

Tropical rain forests are wet environments, but deserts are dry environments.
Tropical rain forests are wet environments. However, deserts are dry environments.

4 Suffixes that change verbs into nouns V

You can change many verbs into nouns by adding a **suffix**, or ending. Two common suffixes that change verbs into nouns are *-ment* and *-ion*/*-ation*. If the verb ends in *-e*, drop the *-e* before adding the suffix.

Look at these examples:

Verb	Noun
erupt	eruption
contribute	contribution
inform	information
argue	argument
assign	assignment

A Look at the verbs. The noun forms of the verbs are in the reading. Find and circle the noun forms. Write each noun form next to the appropriate verb.

Verb	Noun Form	
1. move	____________	(Par. 1)
2. evaporate	____________	(Par. 2)
3. condense	____________	(Par. 3)

Now look at the nouns. The verb forms are in the reading. Find and circle the verb forms. Write each verb form next to the appropriate noun.

Noun	Verb Form	
1. formation	____________	(Par. 3)
2. combination	____________	(Par. 4)
3. continuation	____________	(Par. 4)

B The following sentences need a verb or a noun. Read each sentence. Choose *V* (verb) or *N* (noun) and write the letter in the blank. Then circle the correct word to complete the sentence.

_____ **1.** Our teachers always (*assign / assignment*) a lot of homework.

_____ **2.** Last year I saw a volcanic (*erupt / eruption*). It was amazing!

_____ **3.** In an emergency, listen to the radio for (*inform / information*).

_____ **4.** Don't leave a glass of water in the sun. The water will (*evaporate / evaporation*), and then the glass will be empty.

_____ **5.** The students found the (*locate / location*) of the Sahara Desert on a map.

_____ **6.** The water in a river (*movement / moves*) downhill until it reaches the ocean.

5 Identifying topic sentences W

Many paragraphs in academic English begin with a **topic sentence**, which explains the main idea of the paragraph.

A Look back at paragraphs 2, 3, and 4 of the text "The Water Cycle." Underline the topic sentence of each paragraph. Read the sentences again. Notice that these three topic sentences help you understand the steps of the water cycle.

B Read the paragraph. Then do the following activities.

____ In fact, over half of Earth's plant species live in these very wet environments. We use many of the plants for food or medicine. There are more than 3,000 types of fruit in rain forests, including avocados, coconuts, and guavas. Tropical rain forests also contain a large number of vegetables such as corn, potatoes, and yams, and spices like black pepper, chocolate, and cinnamon. In addition, 70 percent of the plants that can help fight cancer and other diseases are in the rain forest. This rich plant life makes tropical rain forests a very important feature of Earth.

1. What is the main idea of the paragraph? Write it here: ______________________

2. Check (✓) the best topic sentence for the paragraph.

_____ **a.** Tropical rain forests are an important part of the water cycle.

_____ **b.** Many plants and animals live in tropical rain forests.

_____ **c.** Tropical rain forests are home to a wide variety of plants.

PREPARING TO READ

1 Thinking about the topic Ⓡ

Discuss these questions with a partner or in a small group.

1. Freshwater is water that does not have salt in it. Does Earth have more freshwater or saltwater?
2. Freshwater can be ice or liquid. Does Earth have more ice or more liquid water?
3. Where do you get your drinking water?
4. Is your drinking water clean and safe? Do most people have safe drinking water?
5. Will our planet ever run out of safe drinking water?
6. How can we increase the amount of safe drinking water on our planet?
7. How long can a person live without food? How long can a person live without water?

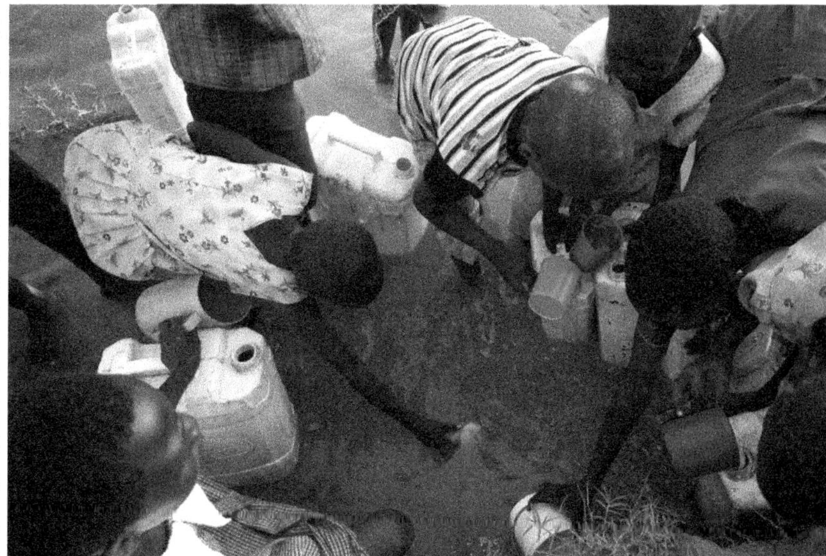

2 Examining graphics Ⓡ

Before reading a text, it is helpful to look at any graphs, charts, or diagrams. This will give you an idea of the content (the information that the text contains).

You are going to read about the distribution of freshwater and saltwater on our planet. Work with a partner. Look at Figure 3.2 on page 62 and answer these questions.

1. Is most of the water on our planet freshwater or saltwater?
2. How much of Earth's water is in the oceans?
3. How much of the freshwater is under the ground (groundwater)?
4. How much of it is in the form of ice?
5. How much freshwater is accessible, or easy to get and use?

Reading 2

GROUNDWATER AND SURFACE WATER

Fresh, clean water is essential to life on Earth. Plants, animals, and humans all need it to survive. People grow plants for food, and that requires a lot of water. In fact, we use four-fifths of the world's freshwater supply for agriculture. To stay healthy, people also have to drink a lot of freshwater every day.

Unfortunately, most of the water on our planet (almost 97 percent) is not freshwater but saltwater. Only about 3 percent is freshwater, and more than three-quarters of that water is frozen, or ice. This means that only a small amount of the water on Earth is freshwater that we can drink. Most of this drinkable water is deep under the ground (groundwater), where we cannot get to it; only about 1 percent is on the surface (surface water) and accessible.

Figure 3.2 The distribution of water on Earth

Groundwater

Inside the lithosphere are trillions of liters of freshwater. Rain and snow fall from the sky, and the water goes down into the ground. It fills the spaces between the sand and the rocks. The water goes farther down and reaches an area full of water. This area is called the zone of saturation. This saturation zone can vary in size. Areas where there is enough groundwater for people to dig a well and pump it out are called **aquifers**. Water in an aquifer may leak out of cracks in the earth and come to the surface, where it goes into lakes and rivers. Most groundwater is too difficult to get or pump out of the earth, however. Out of 33 trillion (33,000 billion) gallons in the ground, only 76 billion gallons are pumped out for use in the U.S.A. – in other words, only 0.23 percent.

Surface water in lakes

Lakes are an important source of accessible freshwater. They provide a large supply of freshwater to the planet. How? First, there are a lot of them. There are millions of lakes on Earth, and almost half of them are in Canada. One very large lake, Lake Baikal in Siberia, contains 20 percent of all the freshwater on the surface of our planet. Another way lakes provide a lot of water is that lake water is easy to get – land surrounds a lake on all sides and makes the water fairly calm and quiet. Last, the freshwater supply in lakes can come from different sources, instead of just one source. Most lakes are in areas that glaciers covered in the past. Some lakes form when rocks stop a river from flowing. Lakes also form when rainwater fills the opening of an extinct volcano.

aquifer an area of rock and soil that holds a lot of water that can be pumped out and used

Surface water in rivers

Rivers are another source of freshwater. Rain and snow fall in the 40 mountains, and some of the water goes into the ground. However, some of the water stays on the surface and flows downhill. This movement often creates a small mountain stream, and this stream may eventually join with other streams to form a river. A young river flows quickly downhill and changes the shape of the land. It cuts a 45 path down the mountain and forms a V-shaped valley. The valley grows larger, the mountain becomes less steep, and the river becomes deeper and wider. Finally, the river reaches flatter land and flows more slowly until it reaches the ocean. The largest river in the world is the Amazon River. In fact, the Amazon contains more than one-fifth of 50 the world's total river water.

Today, concern is growing that Earth's freshwater supply will run out. As the planet's population increases, the global demand for water increases, too. Many people worry that the demand for water could grow higher than the supply. One way to solve the problem may be 55 to find ways to use saltwater. In 2012, a company in California began an experiment to change saltwater from the ocean into freshwater. Another solution is to conserve water, that is, to not use too much. Water is essential to life on Earth. It is extremely important to protect this valuable resource. If we are careful and creative, future generations 60 will also have access to fresh, clean water for everyday use.

Bottled Water or Tap Water?

Where does the water you drink come from? Do you drink tap water from the faucet, or do you drink bottled water? Many people think that bottled water is cleaner and safer than tap water. Surprisingly, this is usually not true. In some parts of the world, the public water supply is polluted, and bottled water is a good choice. In most parts of the world, however, the water supply is clean, safe, and easily accessible.

Bottled water is very popular in several countries. For example, more than 50 percent of Americans, Canadians, and Italians drink it. Why do they drink bottled water? Some people think it is more natural. The labels on the bottles often show mountain streams and beautiful waterfalls, so the water appears to come directly from nature. In fact, more than 25 percent of all bottled water that is sold in the United States is purified water, and this water comes from a faucet, not a mountain stream. Other people think bottled water tastes better than tap water. However, some tests say that most people cannot taste the difference. Therefore, if you do have access to a good water supply, you may want to turn on the faucet for your next drink. You'll save money, and the water will be clean, safe, and good tasting.

AFTER YOU READ

1 Answering multiple-choice questions A R

Students often focus on the choices in multiple-choice questions, but it's important to think about the questions. Pay special attention to questions that include the word *not*.

Example: Which one of the following is not a step in the water cycle?

a. evaporation
b. saturation
c. condensation

In the example, choices *a* and *c* (*evaporation* and *condensation*) are steps in the water cycle. The correct answer is *b*, because *saturation* is not a step.

A Review "Answering Multiple Choice Questions" on page 21. Use the information in the reading to answer the following questions. Pay special attention to questions with *not*.

1. How much of the freshwater on Earth is accessible surface water that we can drink?

a. about 97 percent
b. about 75 percent
c. about 3 percent
d. about 1 percent

2. Most of the freshwater on our planet is ____.

a. under the ground
b. frozen
c. in oceans
d. in lakes

3. An aquifer is ____.

a. under the ground
b. on the surface
c. very small
d. sand and rocks

4. Which one of the following is not an example of surface water?

a. a river
b. a lake
c. an aquifer
d. an ocean

5. Which phrase does not describe lakes?

a. freshwater
b. surrounded by land
c. calm water
d. flow from mountains to the ocean

6. Rivers do not ____.

a. flow downhill
b. change the shape of the land
c. create U-shaped valleys
d. flow into oceans

B Compare your answers with a partner.

2 Mapping Ⓐ Ⓡ

> Make a map of a text to take notes. One way to do this is to draw lines and circles to show relationships among the facts and ideas.

A Look at the illustration below. It is the beginning of a map of "Groundwater and Surface Water." The circle in the center is large and its content is general. The other circles become smaller as the content becomes more specific.

B Complete the map. Fill in the blank circles with information from the text. Then compare your map with a partner.

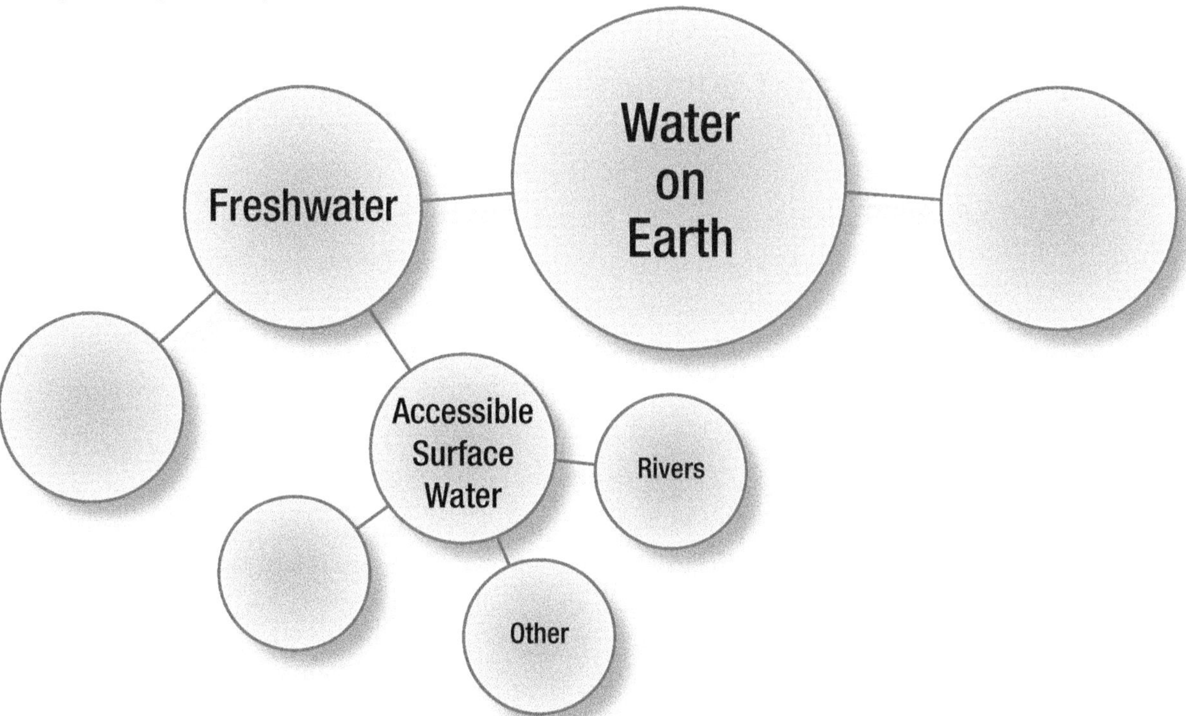

3 Conducting a survey Ⓐ

> Surveys are often used in academic research. Two types of surveys are written questionnaires and interviews. In questionnaires, people write answers to questions on paper. In interviews, a person asks questions and then writes down people's answers. A good survey does the following:
>
> - poses a clear question
> - often asks for comparisons: *Is X better than Y?*; *Is X different from Y?*
> - uses yes / no and multiple-choice questions
>
> Note that yes / no and multiple-choice questions provide information that is easy to calculate. When the survey is complete, researchers usually present the results in a report or chart.

A Conduct a survey to answer this question: Can most people tell the difference between bottled water and tap water?

1. Ask at least six people to participate in your survey.
2. Prepare four cups of water for each person. Fill three cups with bottled water and one cup with tap water. Label the cups *A*, *B*, *C*, and *D*. Be sure to secretly note which cup contains the tap water.
3. Do not tell your participants what the cups contain – this is called a "blind" taste test.
4. Have each person drink from the four cups. Ask each person, "Which cup do you think contains tap water?" Record their names and answers on a separate piece of paper.

B How many people correctly identified the tap water? How many did not? Calculate the numbers. Make a pie chart to show the results. Share the results with the class.

4 Countable and uncountable nouns 

English nouns can be divided into two groups:

1. **Countable nouns** are things that we can count.
   - Countable nouns have both singular and plural forms: *river / rivers*
   - Countable nouns can take a determiner: *a*, *an*, *the*, or a number

     A river is a source of freshwater.

     The longest **river** in Canada is the Mackenzie River.

     Two rivers in Russia are the Volga River and the Don River.
2. **Uncountable nouns** are things that cannot be counted.
   - Uncountable nouns are often:

Liquids	**Gases**	**Ideas**	**Things made up of little pieces**
water	oxygen	love	rice
coffee	carbon dioxide	peace	dust

   - Uncountable nouns do not have plural forms. They take singular verbs.

     Water is essential for life on Earth.

     When **water condenses**, it forms clouds.
   - Uncountable nouns do not take *a*, *an*, or a number. You can use the noun alone or with *the*.

     CORRECT: **Water** is dripping from the faucet.

     The water in a lake is freshwater.

     INCORRECT: A water is good for your health.

Note: Do not confuse a plural countable noun with a third-person singular verb:

I have a pair of *skis*. (plural countable noun) He *skis* on the river. (verb)

A Reread paragraph 3 of "Groundwater and Surface Water." Find three countable nouns and three uncountable nouns. Circle them. Write the words in the blanks.

Countable nouns: ________ ________ ________

Uncountable nouns: ________ ________ ________

B Read the sentences. Label each underlined noun *C* (countable) or *U* (uncountable).

1. The United States is between two oceans: the Atlantic Ocean and the Pacific Ocean.
2. When Mount Vesuvius erupted, lava covered the city of Pompeii, Italy.
3. Most of Antarctica has a thick cover of ice.
4. The Nile River and the Amazon River are two of the longest rivers on Earth.
5. An example of an aquifer is the Ogallala Aquifer in the midwestern part of the United States.

5 Reading about statistics Ⓡ

Statistics are numerical facts, or facts that are stated as numbers. Statistics are often used to support ideas. Understanding the statistics in an academic text will help you understand the ideas. Statistics can be stated in different forms:

- whole numbers (50, 1,000, 32 billion)
- fractions (one-third, one-half, four-fifths, 1/3, 1/2, 4/5)
- percentages (33%, 50%, 80%)

A Read the questions and answer with a partner. Use the statistics in the main reading and the boxed text.

1. Do people use more water for agriculture or for drinking?
2. What percentage of water on Earth is freshwater?
3. Is most freshwater in liquid form or frozen?
4. Does all bottled water come from natural sources, such as streams?

B Statistics stated in a text may be stated differently in questions. For example:

Text: **Fifty percent** of people in Canada drink bottled water.

Question: **What fraction** of people in Canada drink bottled water?

Answer: one-half (1/2)

Match the fractions with the percentages.

____ 80 percent	**a.** one-fourth (1/4)
____ 50 percent	**b.** one-tenth (1/10)
____ 33 percent	**c.** four-fifths (4/5)
____ 20 percent	**d.** one-fifth (1/5)
____ 25 percent	**e.** one-third (1/3)
____ 10 percent	**f.** one-half (1/2)

C Answer the questions. Use the statistics in the reading and the boxed text.

____ **1.** What percentage of the world's freshwater supply is used for agriculture?

____ **2.** What fraction of Earth's surface freshwater is found in Lake Baikal?

____ **3.** What percentage of the world's lakes is in Canada?

____ **4.** What fraction of bottled water in the United States is purified water?

D Compare your answers to steps A–C with a partner's.

PREPARING TO READ

Increasing reading speed Ⓡ

Academic classes often require a lot of reading. However, there is not always time to read every text slowly and carefully. Reading speed can be as important as reading comprehension. Here are some strategies for increasing your reading speed:

- Read the text straight through. Do not go back to any parts of it.
- Do not stop to look up words in a dictionary.
- Skip over words you do not know if they do not seem important.
- Try to guess the meaning of words that seem important.
- Slow down a little to understand important parts, such as definitions and main ideas.

A Read the text "Glaciers." Use the strategies listed above. For this task, do not read the boxed text on page 70.

1. Before you begin, fill in your starting time.
2. Fill in the time you finished.

Starting time: ________

Finishing time: ________

B Calculate your reading speed:

Number of words in the text (418) ÷ Number of minutes it took you to read the text = your reading speed

Reading speed: ________

Your reading speed = the number of words you can read per minute.

C Check your reading comprehension. Circle the correct answers. Do not look at the text.

1. Glaciers are made of
a. freshwater and saltwater.
b. soil and rocks.
c. layers of ice.

2. Glaciers can move
a. rocks and soil.
b. people and animals.
c. rivers and oceans.

3. Glaciers move
a. quickly.
b. slowly.
c. not at all.

Reading 3

GLACIERS

In a small town in Alberta, Canada, there is a giant rock called Big Rock. This rock is 9 meters high, 41 meters long, and 18 meters wide. It weighs almost 14,970 metric tons. Thousands of years ago, Big Rock moved approximately 400 kilometers to its present location. How could such a huge rock move so far? It got a ride from a glacier.

Glaciers are layers of ice that move. Some people call them rivers of ice. Glaciers begin to form when snow falls and does not melt. More snow falls, presses down, and turns the snow underneath to ice. This cycle eventually forms thick layers of ice. The cycle continues until the layers of ice become very heavy and begin to slide over the ground. When they begin moving, the layers of ice are called a glacier. They slide slowly down hills and valleys until they melt or reach the ocean.

Glaciers may move slowly, but they are powerful. They can carve, or cut out, deep valleys and create beautiful landscapes over time. They are very strong and can move large amounts of soil and pull rocks of all sizes out of the ground. This is called *plucking*. Several glaciers can surround a mountain and pluck rocks from all sides. This is the process that created the sharp mountaintop of the Matterhorn in Switzerland.

fjord valley near the ocean, originally created by a glacier, that has filled with ocean water

Glaciers are usually wider than rivers. They create wide U-shaped valleys, not narrow V-shaped ones, as they move across the land. Sometimes a glacier will carve a U-shaped valley near the ocean. After the glacier melts, the ocean water sometimes fills these valleys. This creates a landform called a **fjord** (fee-YORD). Norway has the world's most famous and beautiful fjords, but there are also fjords in Alaska and Japan.

A glacier in the Alps, Switzerland

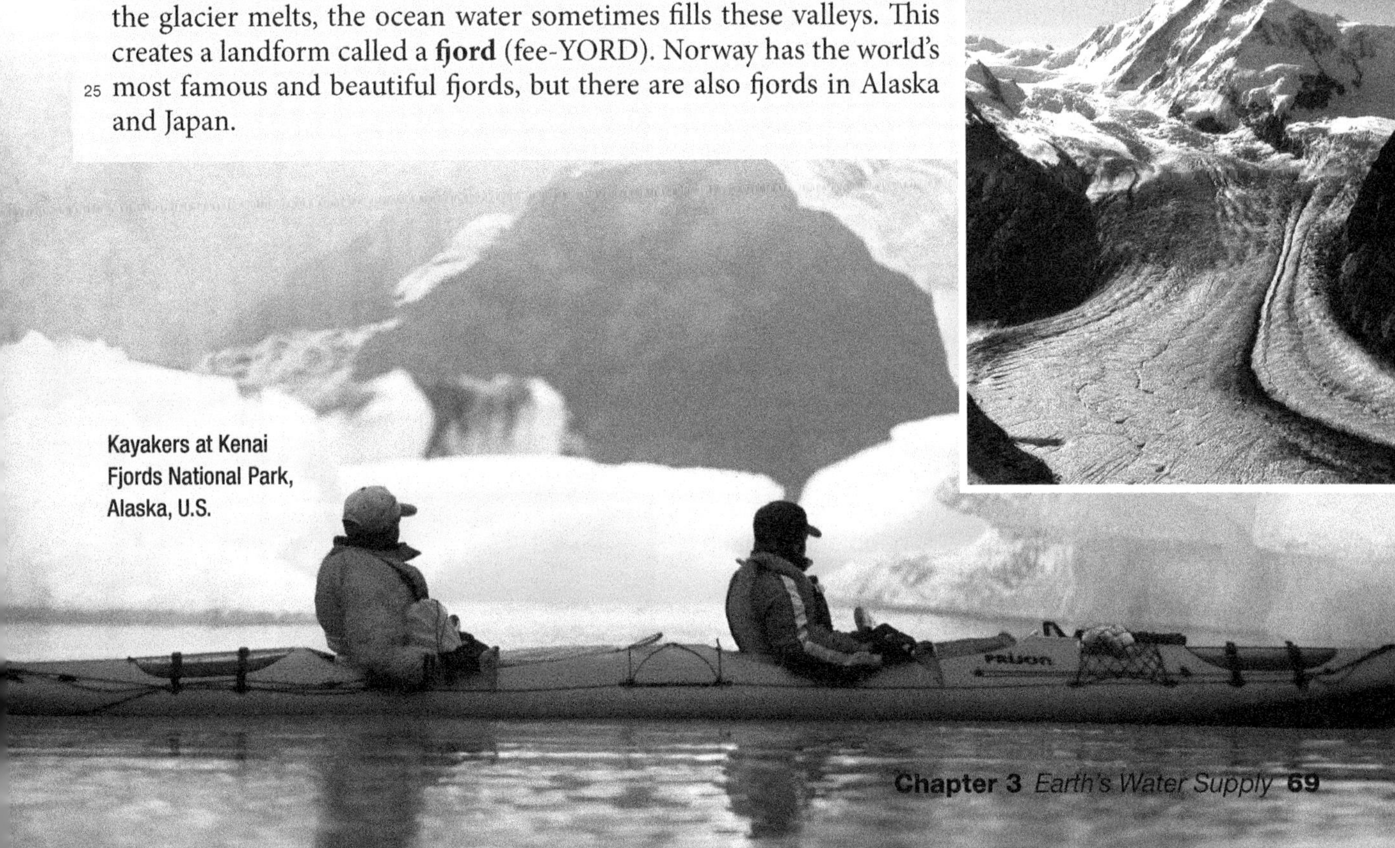

Kayakers at Kenai Fjords National Park, Alaska, U.S.

climate the usual weather conditions in a particular place

Scientists study glaciers to learn about world **climates** and climate change. They measure how fast glaciers move, and they measure changes in size. Today, scientists report that Earth's glaciers are melting 30 more quickly than they melted 100 years ago. This is an important sign that Earth is warming up.

Glaciers are also important because they contain more than 75 percent of all the freshwater on our planet. When they melt, they provide drinking water and water for crops in some dry parts of the world. 35 However, some glaciers are melting too quickly. They are decreasing in size, and eventually they will melt and disappear. As a result, many people will lose their main source of water. Scientists and environmentalists all over the world are trying to solve this serious problem.

Artificial Glaciers

Sometimes, nature makes life difficult. For example, farmers in Ladakh, India, need water in April to prepare the fields and plant seeds, but there is no water in Ladakh in April. Melting glaciers from the Himalayan mountains provide water to grow crops, but the glaciers are high in the mountains and don't begin to melt until June. Unfortunately, the farmers need the water in April, and June is too late to have a good harvest. As a result, the people in this area are very poor and must work hard to survive.

One person, however, can change the lives of thousands. Chewang Norphel lived his whole life in Ladakh. He worked as an engineer and learned to build bridges and canals. He understood that the farmers needed more water in April, and he found a way to solve their problem. Norphel built a system of canals to catch the water that flows in the glacial streams in the summertime. They store this water in reservoirs until winter. In the winter, the water refreezes and becomes an artificial, or human-made, glacier.

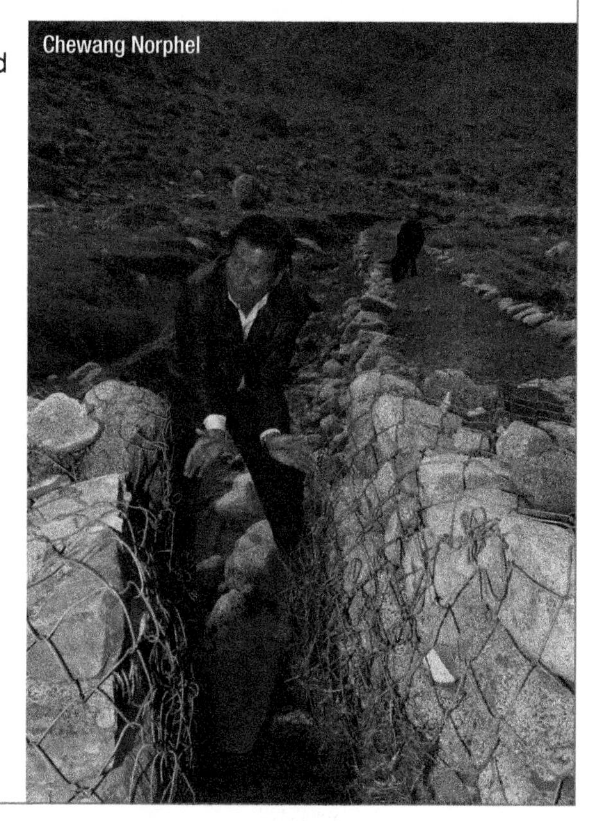
Chewang Norphel

The artificial glaciers are lower in the mountains than natural glaciers, so they melt earlier. In fact, they melt in April. This provides water for the farmers just when they need it. Norphel has built 12 artificial glaciers. He hopes to create many more. Today, the people in Ladakh can grow more food, and their living situation is improving.

AFTER YOU READ

1 Reading for main ideas (R)

A Look at the main ideas of "Glaciers." Find the paragraphs in the reading where they are discussed. Write the numbers in the blanks.

1. Glaciers create U-shaped valleys, and some of these eventually become fjords. Par. ____
2. Glaciers are large, slow-moving layers of ice. Par. ____
3. Studying glaciers can provide information about world climates. Par. ____
4. Glaciers can move huge rocks hundreds of kilometers. Par. ____
5. Glaciers can create valleys, pull rocks out of the ground, and shape mountains. Par. ____

B Check (✓) the sentence that expresses the main idea of the whole text.

____ **1.** Glaciers are very powerful, and they can lift heavy objects.
____ **2.** Glaciers are melting quickly, and we need to find a way to save them.
____ **3.** Glaciers are layers of ice that shape the land, and they can teach us about climate.
____ **4.** Glaciers are layers of ice that move slowly over the land.

2 Scanning (R) (A)

Scanning a text means reading it quickly to find specific information. To scan, you do not read every word. Instead, you pass your eyes quickly over the text to find only the information you need. You can use scanning to study for a test or prepare for a writing assignment.

Scan for:

- names of people, places, or things
- facts or statistics
- signal words that show order or examples

Search the text for:

- capital letters, to find names of people or special places
- numbers or percent symbols (%), to find amounts or measurements
- words such as *first*, *second*, *last*, or *for example* to find signals

Scan the text "Glaciers" and the boxed text "Artificial Glaciers" to find the information.

1. Where is Big Rock located? __________
2. How long is Big Rock? __________
3. How many kilometers did Big Rock move? __________
4. What is the shape of a valley that a glacier creates? __________
5. What are three places that have fjords? __________
6. Where is Ladakh? __________
7. How many glaciers has Chewang Norphel made? __________

3 Subject-verb agreement

Singular subjects take singular verbs. Remember that, in the present tense, third-person singular verbs end in *-s* or *-es*.

subject	verb	
Big Rock	**weighs**	almost 14,970 metric tons. **It is** 9 meters high.
subject	verb	
A glacier	**carves**	deep valleys as **it slides** over the ground.

Plural subjects take plural verbs. Plural verbs do not end in *-s*.

subject	verb	
Scientists	**study**	glaciers to learn about world climates.

Uncountable nouns take singular verbs.

subject	verb	
Water	**covers**	97 percent of Earth's surface.

Complete the sentences with the correct form of the verb in parentheses.

1. A glacier __________ (form) when the snow in an area does not melt. Over time, this snow __________ (turn) into ice.
2. In some glaciers, the ice __________ (be) thousands of years old.
3. Glaciers __________ (contain) more than 75 percent of Earth's freshwater.
4. Every continent __________ (have) glaciers except Australia.

4 Identifying topic sentences and supporting sentences 

Many paragraphs in academic English begin with a **topic sentence**. **Supporting sentences** follow the topic sentence.

- **topic sentences:** state the main idea of a text
- **major supporting sentences:** directly support a topic sentence; answer questions such as *how*, *why*, or *when*
- **minor supporting sentences:** directly support a major supporting sentence; answer questions such as *how*, *why*, or *when* with specific information

A Read the paragraph below.

Glaciers change the surface of our planet in different ways. One way is by shaping the land. For example, glaciers carve U-shaped valleys and form sharp mountaintops. Glaciers can also move big rocks to other locations. Another way glaciers change Earth is by creating lakes. For instance, Mirror Lake and the Great Lakes in the United States were formed by glaciers. Lake Louise in Canada is another example. Earth would look very different without the work of glaciers.

B Highlight the topic sentence.

C Highlight the sentences that provide major support for the topic sentence. Use a second color.

D Highlight the sentences that provide minor (or indirect) support for the topic sentence. Use a third color.

E Notice that the last sentence in the paragraph is the concluding sentence.

F Answer *T* for true or *F* for false. Compare answers with a partner.

_____ **1.** The sentence "One way is by shaping the land" directly supports the main idea.
_____ **2.** The paragraph gives examples to prove the major supporting sentences.
_____ **3.** The main idea of the paragraph is that glaciers can carve the land.

5 Writing topic sentences and supporting sentences W

A Look at the topics. Write a topic sentence for each topic. Remember: Topic sentences state main ideas that can be proved. Examples, illustrations, facts, or statistics show or prove your point.

Example:

Topic: Glaciers and climate

Topic sentence: *Glaciers can tell us about climate.*

Major support: *For example, glaciers can show signs of global warming.*

Minor support: *Glaciers that melt fast can show that Earth is getting hot.*

1. Topic: Lakes and sources of freshwater

Topic sentence: ______________________________

Major support: ______________________________

Minor support: ______________________________

2. Topic: The importance of glaciers

Topic sentence: ______________________________

Major support: ______________________________

Minor support: ______________________________

B Go back to Step A. Write one major supporting sentence and one minor supporting sentence for each topic.

Chapter 3 Academic Vocabulary Review

The following words appear in the readings in Chapter 3. They all come from the Academic Word List, a list of words that researchers have discovered occur frequently in many different types of academic texts. For a complete list of all the Academic Word List words in this chapter and in all the readings in this book, see the Appendix on page 206.

accessible	energy	generations	occur
approximately	environment	global	percent
contrast	environmentalists	labels (n)	source

Complete the following sentences with words from the list above.

1. In 2012, about half of the people in the United States drank soda every day, and 64 ________ drank coffee every day.
2. Air and water pollution is a ________ problem. People from many different countries struggle with this issue.
3. Some farmers use wind ________ to pump water from the ground.
4. Chewang Norphel's family has lived in India for many ________.
5. Oceans and rivers are a major ________ of income for people who fish their waters.
6. Most dolphins live in the ocean, but the pink river dolphin lives in a freshwater ________.
7. Mount Kosciuszko in Australia is considered an ________ mountain, and more than 100,000 people climb it every year.
8. Many people enjoy hiking in the mountains, but problems can ________. Do not hike in the mountains unless you are familiar with the area.
9. ________ often warn us that we might destroy our planet. They say we must not pollute the water, land, and air.
10. Today, glaciers cover about 10 percent of the land on Earth. This is a huge ________ to twenty thousand years ago; then, glaciers covered 32 percent of the land on Earth.

Developing Writing Skills

In this section, you will practice writing topic sentences and two types of supporting sentences. You will also use what you learn here to complete the writing assignment at the end of this unit.

Types of Major and Minor Support

Remember that the topic sentence states the main idea and that supporting sentences follow the topic sentence. Here is more information about types of support:

Major support (major details):

- directly support or explain the topic sentence
- can include major examples and illustrations

Minor support (minor details):

- directly support or explain the major supporting sentences
- can include examples, illustrations, facts, and statistics
- develop the major details and make the main ideas stronger and more convincing

A Read the paragraph. Think about the types of ideas: main, major, and minor.

____.

The first type of desert is the hot desert. For instance, the Mojave and the Sahara are both hot deserts. These deserts have high temperatures in the daytime, cooler temperatures at night, and just a little rain. Their average temperatures are 20°–25°C and most receive less than 15 centimeters of rain each year. Only a few plants, such as prickly pears and acacias, can live in hot deserts. The second type of desert is the cold desert. The Gobi and Namib are both examples of cold deserts. Some cold deserts have high temperatures in the summer, but very cold temperatures in the winter. The average winter temperatures are -2° to 4°C. Cold deserts have almost no rain, but some snow. On average, they receive 15–26 centimeters of snow each year. There are only a few plants, such as sagebrush, in cold deserts. These differences clearly show that all deserts may be dry, but they are not all the same.

B Work with a partner. Find the support in the paragraph. Circle the major details. Underline the minor details. Then answer the following questions.

1. How many major details are there? ______________
2. What signal words introduce the major details? ______________
3. What are the types of minor details? ______________

C Now think about the main idea. The topic sentence of the paragraph is missing. Write one in the blank. Compare your sentence with your partner's.

D Read the following pieces of information. Think about major and minor details. Read the descriptions of the types of details in the box above again if you need to.

- One leaky faucet can waste more than seven liters of water each day.
- Many people believe that the Nile River is the longest river in the world.
- Try to take shorter showers.
- It is more than 6,500 kilometers long and flows through nine countries, including Egypt, Kenya, and Tanzania.
- For example, if you take a 5-minute shower instead of a 10-minute shower, you could save almost 100 liters of water.
- The Yangtze River is the most famous river in China.
- Check the faucets in your home for leaks, and fix any problems.
- It is about 4,990 kilometers long and divides northern China and southern China.

E Work with your partner. Sort the sentences above into two groups (Set A and Set B), by topic. Then sort the sentences again by type of detail. Write the sentences in the chart below.

Set A
Main idea: ______________________________
• *Check the faucets in your home for leaks, and fix any problems.*
•
•
•
Set B
Main idea: ______________________________
•
•
•
•

F Now look at the details above. What is the main idea of each set of details? Write the main idea for Set A and for Set B. Compare your answers with a partner.

G Imagine that you must write a paragraph on this topic: *How do rivers and lakes affect our lives?* Work in a small group. Discuss the information in this chapter and your own ideas. Write a topic sentence. Write details to support it. Remember: minor details make your ideas convincing. Share your ideas with the class.

Chapter 4
Earth's Oceans

PREPARING TO READ

1 Thinking about the topic Ⓡ

You are going to read about the oceans on Earth. How much do you know about this topic? Discuss the following questions in a small group.

1. How many oceans are there? Name the ones you know.
2. How are oceans different from rivers and lakes?
3. How does ocean water taste?
4. What are some living things that you can find in the ocean?
5. Why are oceans an important part of life on our planet?
6. What problems connected to oceans do people worry about these days?

2 Building background knowledge about the topic Ⓡ

The word *coast* refers to land that is next to or close to an ocean. Read the list of facts about coastal areas. Then discuss the questions with your classmates.

Facts About Coastal Areas

- More than half of the people on Earth live less than 100 kilometers from an ocean.
- Almost two-thirds of the world's largest cities are in coastal areas.
- Since 1970, almost 50 percent of the construction of new homes and businesses in the United States has been in areas along the coasts.

1. Why do so many people live near an ocean?
2. What are some advantages of living near an ocean?
3. How do large populations who live near an ocean affect the ocean's water and animals?
4. How do high water levels of the ocean affect people's lives?

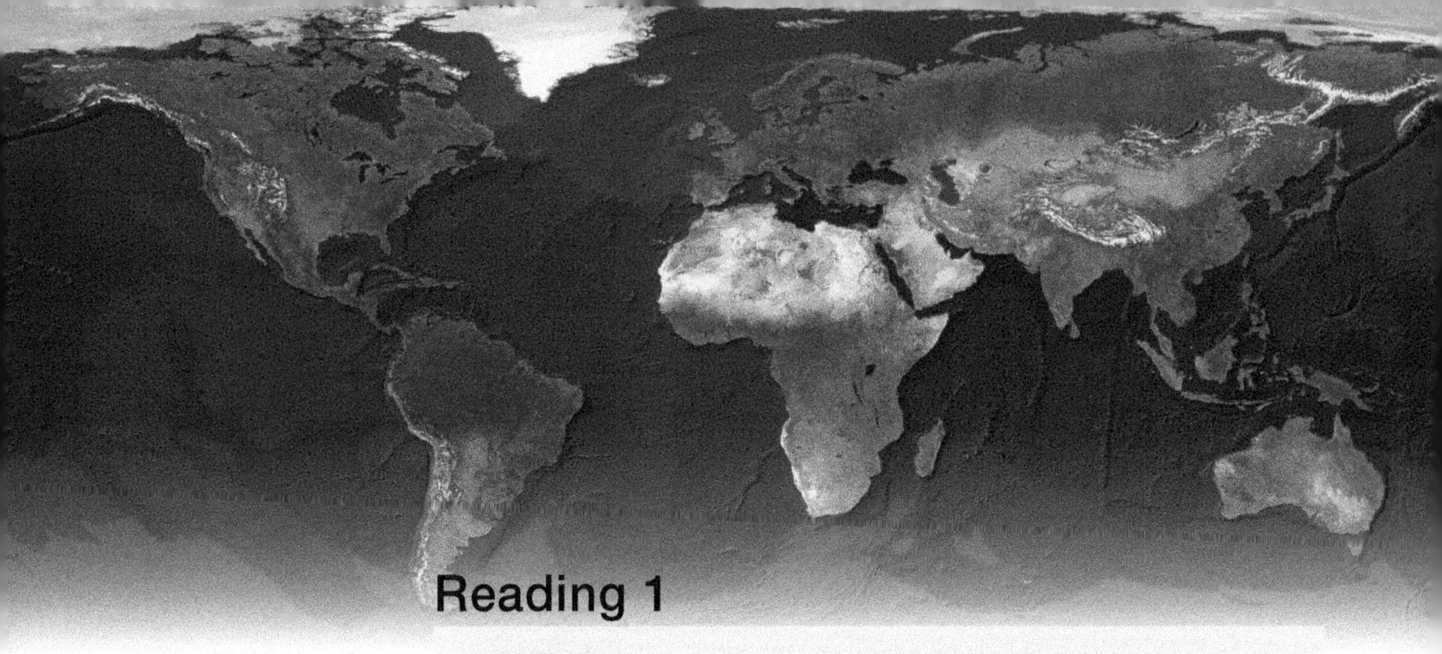

Reading 1

OCEANS

One nickname for Earth is "the blue planet." From outer space, all the ocean water makes the planet look mostly blue. There are four main oceans on Earth: the Pacific, the Atlantic, the Indian, and the Arctic. Many scientists include the Southern Ocean, also called the 5 Antarctic Ocean, as a fifth ocean. Together, the oceans cover more than 70 percent of Earth's surface, and they flow into each other. Therefore, from outer space it looks as if Earth has one huge blue ocean.

The main oceans

The Pacific is the largest and deepest of the main oceans. It contains approximately half of the ocean water on our planet. The first time 10 that the explorer Ferdinand Magellan sailed on this huge ocean, it was a calm day. That's why Magellan named the ocean *Mar Pacífico* – that is, "peaceful ocean" or "calm ocean" in Magellan's native language of Portuguese. Today, however, we know that the Pacific Ocean is not always peaceful. There are frequent earthquakes and volcanic 15 eruptions.

The Atlantic Ocean is the second-largest ocean. It covers about a fifth of Earth's surface. The Indian Ocean is sometimes called the calmest ocean. It is a little smaller than the Atlantic. The Arctic Ocean is Earth's smallest and coldest ocean.

The salinity of oceans

salinity a measure of the amount of salt in ocean water

20 Ocean water is salty. You cannot drink the water; you'll notice the salt right away. Ocean water is about 96.5 percent water and 3.5 percent salt. The **salinity** of an ocean varies, depending on two main factors: the amount of evaporation and the amount of freshwater that is added. Ocean water evaporates and leaves salt behind. Oceans that have a 25 lot of evaporation are saltier than oceans that have less evaporation. Rivers or rain bring freshwater to an ocean, and this makes the ocean less salty. The freshwater dilutes, or weakens, the salt.

In areas near the **equator**, oceans are very salty. The heat of the sun causes a lot of evaporation. There is not a lot of rain, so there are 30 high levels of salinity in the warm ocean water. In cold areas near the **North Pole** and the **South Pole**, oceans are not as salty. There is less evaporation, and the oceans receive freshwater from melting glaciers. Therefore, salinity is lower in polar areas. The lowest salinity levels occur where large rivers empty into an ocean. That's why the place 35 where the giant Amazon River flows into the Atlantic Ocean is less salty than the rest of the ocean.

the equator an imaginary line around Earth that divides it into two equal halves: the Northern and Southern Hemispheres

North Pole Earth's northernmost point, which is located in the Arctic Ocean

South Pole Earth's southernmost point, which is located in Antarctica

The role of oceans in our lives

Oceans do not provide our drinking water, but they affect us in many ways. First, oceans are an important part of the water cycle. Most of the evaporation on Earth is from ocean water. Oceans also 40 provide us with food and jobs. Millions of people work in the fishing, shipbuilding, offshore oil and gas industries, international shipping and trade, and ocean science. In addition, the ocean is home to many of our planet's plants and animals. Finally, coastal areas are popular places to live. Today, about 50 percent of the world's people live close 45 to an ocean.

Dirty Water

Oceans are an essential part of life on our planet, but unfortunately we don't always treat them with respect. Many of us use the ocean as a trash can. Some areas of the ocean look like garbage dumps, with plastic bottles, old fishing nets, tires, straws, cigarettes, plastic bags, and other trash floating in the water. Every ocean has problems with marine debris. In the summer of 2012, scientists from the National Oceanic and Atmospheric Administration (NOAA) removed 50 metric tons of garbage from the Pacific Ocean waters and the shorelines of the Northwestern Hawaii Islands.

Marine debris is very dangerous. Each year, the trash kills thousands of ocean animals. Old fishing nets trap seals, birds, whales, fish, and turtles. Animals often think the pieces of garbage, especially plastic pieces, are food. For instance, turtles like to eat jellyfish. To turtles, plastic bags look like jellyfish, so they eat them. Then, they get sick and sometimes die. Plastic has been found in seabirds, turtles, fish, and whales. In fact, scientists found a bird that had swallowed more than 400 pieces of plastic, and a gray whale with plastic bags, surgical gloves, rope, fishing line, duct tape, towels, and a golf ball in its stomach.

It is clear that marine debris is a serious problem in many places. However, a few simple changes can improve the health of Earth's oceans. We can replace plastic bottles and bags with reusable ones, and this will decrease the amount of plastic that goes into the oceans. We can decide not to litter. We can also participate in neighborhood cleanups. Small changes can have a big effect on the oceans.

AFTER YOU READ

1 Taking notes Ⓐ

Learning to take good notes is very important. Good notes can help you remember and review a text you have read. People usually have their own ways of taking notes, but everyone should follow these basic guidelines:

- Be sure to include all the important ideas and examples.
- Organize the notes in a logical way.
- Write only important words, not complete sentences.
- Use abbreviations and symbols.

Look at this page from a student's notebook. It shows the beginning of the student's notes on the text "Oceans." Use the information in the reading to complete the notes. Then compare your notes with a partner's.

Oceans

General info

5 oceans: Pacific, ________________ , ________________ ,
Arctic, Southern

Cover ______ % of Earth's surface

Main oceans & features

Pacific: largest, ________________ , often violent

Atlantic: 2nd ________________ , covers ________________ of
Earth's surface

Indian: calmest, ________________ than Atlantic

Salinity (= saltiness)

Ocean water = 96.5% water + ______ salt

Depends on: 1) amount of ________________
2) amount of freshwater

Higher near the equator, lower near the ________________ & the
places where ____________________________________

Importance of oceans

Ex: 1. Role in water cycle
2. Provide ________________
3. Provide jobs
4. Home for many ________________ + ________________
5. People like to live nearby

2 Reading maps R A

Most maps have a key that helps you read them. The key explains the marks and symbols used on the map.

A Study the map below. It shows general patterns of ocean salinity.* Read the key. What do the three colors show?

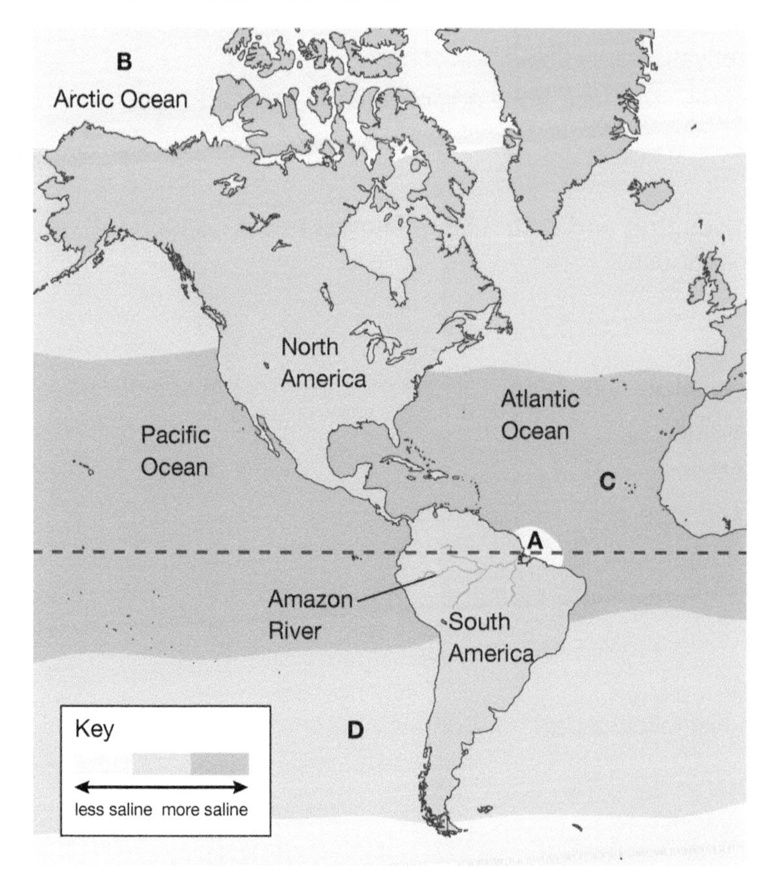

*Note that salinity levels vary within these general patterns.

B Find these areas on the map in Step A. Use the map and the key to decide the level of salinity of each area. Write *L* (low), *M* (medium), or *H* (high).

____ **1.** area A

____ **2.** area B

____ **3.** area C

____ **4.** area D

C Compare your answers with a partner.

3 Writing about superlatives (W)

Academic texts often compare or describe results. A **superlative adjective** is used to compare three or more people, places, or things in a group. Superlatives begin with *the*. Look at these examples:

The Pacific is **the deepest** ocean in the world. It is deeper than all the other oceans.

Some people think Columbus is **the most famous** explorer. They think he is more famous than Magellan or Vasco da Gama.

To form superlative adjectives, follow these guidelines:

For **one-syllable adjectives**, add *-est*. If the adjective ends in *-e*, add *-st*.

deep ➔ the deepest

wide ➔ the widest

For **one-syllable adjectives that end with a single vowel and a consonant**, double the final consonant and add *-est*.

hot ➔ the hottest

big ➔ the biggest

For adjectives **with two or more syllables**, add *the most* before the adjective.

careful ➔ the most careful

patient ➔ the most patient

dangerous ➔ the most dangerous

interesting ➔ the most interesting

If an adjective ends in *-y*, change the *-y* to *-i* and add *-est*.

rocky ➔ the rockiest

cloudy ➔ the cloudiest

Irregular adjectives do not follow a pattern. Some common ones are:

good ➔ the best

bad ➔ the worst

far ➔ the farthest

A Reread the text "Oceans." Circle all of the superlative adjectives. How many did you find?

B Write the superlative form of each adjective below.

1. large ____________________

2. cold ____________________

3. small ____________________

4. salty ____________________

5. peaceful ____________________

6. important ____________________

C Complete the sentences with the superlative form of an adjective in the box. One adjective is not used.

calm	~~deep~~	small	successful	violent

1. The Pacific is the deepest ocean in the world. It is deeper than all the other oceans.
2. The Pacific might also be __________ ocean. It has many earthquakes and volcanoes.
3. The Indian Ocean might be ____________________ ocean. The water is usually quiet.
4. The Arctic Ocean is ________________________ ocean. All the other oceans are larger.

D Work with a partner. On a separate piece of paper, write about the ocean. You could write about a specific ocean, plant or animal life in the ocean, or people's interactions with the ocean. Write four or five sentences and use a superlative adjective in each sentence.

Example: *The most popular ocean sport in Hawaii is surfing.*

4 Describing results W

A sentence beginning with *therefore* or *that's why* explains the effect or result of something. The preceding sentence (the sentence that comes before) tells the cause of the result or effect. Look at these examples:

cause	effect (result)
Ocean water contains about 3.5 percent salt.	**Therefore,** it tastes salty.

cause	effect (result)
Ocean water contains about 3.5 percent salt.	**That's why** it tastes salty.

Notice that *therefore* is followed by a comma.

A Reread the text "Oceans." Find the sentences with *therefore* and *that's why*. Underline them. Notice the cause-and-effect relationships. Look at the sentence that comes before *therefore* or *that's why*. Does it state a cause?

B Study the examples in the box above. Then work with a partner. Match the effects and causes. Make pairs of sentences using *therefore* and *that's why*.

Cause	Effect (Result)
____ **1.** The five oceans flow into each other.	**a.** He named the ocean *Mar Pacífico*, which means "peaceful ocean."
____ **2.** When Magellan first sailed on the Pacific, it was calm.	**b.** From outer space, it looks as if Earth has one huge ocean.
____ **3.** In areas near the equator, there is a lot of evaporation and not a lot of rain.	**c.** The place where the Amazon River empties into the Atlantic Ocean is less salty than the rest of the ocean.
____ **4.** Fresh river water dilutes the salt in ocean water.	**d.** Ocean water near the equator usually has higher levels of salinity.

5 Concluding sentences W

Remember that many paragraphs in academic English end with a **concluding sentence**. The concluding sentence restates the main idea of the paragraph in different words.

A Look back at paragraph 1 of "Oceans." Label the concluding sentence (*CS*). Compare the concluding sentence with the topic sentence.

B Read the following paragraph. It is missing a concluding sentence.

The Arctic and the Southern Oceans differ in several ways. The Arctic Ocean is located at the North Pole, and it is surrounded by Canada, Norway, Greenland, Russia, and the United States. The Arctic Ocean is approximately 12 million square kilometers, and it is Earth's smallest ocean. It is also the coldest ocean. Its surface is often frozen in the winter, and much of its ice never melts. The Southern Ocean is at the South Pole. The Southern Ocean surrounds Antarctica. It is more than 20 million square kilometers and, therefore, larger than the Arctic Ocean. It is also warmer than the Arctic Ocean. Temperatures in the Southern Ocean range from -2° to 10°C (28° to 50°F), and strong winds often blow across its surface. In fact, the Southern Ocean is the windiest ocean on Earth. ______________________________

Check (✓) the best concluding sentence for the paragraph above.

_____ **1.** Therefore, the Arctic Ocean is smaller than the Southern Ocean.

_____ **2.** Clearly, the Arctic and Southern Oceans are important to life on our planet.

_____ **3.** The facts clearly show that the Arctic and Southern Oceans are different in location, size, and weather.

C Read the following paragraph. It is missing a concluding sentence. Write one.

Jacques-Yves Cousteau (1910–1997) was a French explorer. Cousteau dedicated his life to understanding and protecting the oceans. As a young man, he joined the French navy, and he began to do underwater research. He and an engineer, Emile Gagnan, created the Aqua-Lung. The Aqua-Lung is a piece of scuba-diving equipment that allows people to stay underwater for long periods of time. For more than 40 years, Cousteau and his crew explored Earth's oceans and conducted research on his ship, the *Calypso*. He shared his knowledge and love of the oceans through his television films and books, and he helped people understand the need to protect this valuable resource.

____

______________________________.

Jacques-Yves Cousteau

PREPARING TO READ

1 Thinking about the topic Ⓡ

A The text you are going to read is about ocean currents. Read this description of a current:

A current is like a river of warm or cold water that flows through the ocean.

B Discuss these questions in a small group.

1. What do you think causes ocean currents?
2. Do you know the names of any currents?
3. Why are currents important?
4. What is a rip current? Make a guess.

2 Examining graphics Ⓡ

Work with a partner. Look at Figure 4.1 on page 86. It shows several currents on the surface of the oceans. Also, look at the drawing of a rip current on page 87. Then read the statements below and write *T* (true) or *F* (false).

_____ **1.** Near the equator, most currents flow from east to west.

_____ **2.** Farther away from the equator, most currents flow from east to west.

_____ **3.** In general, currents flow in a circle.

_____ **4.** All currents are warm currents.

_____ **5.** In the Atlantic Ocean, there is a warm current that flows along the east coast of North America.

_____ **6.** Rip currents can be dangerous for swimmers.

_____ **7.** Rip currents are very wide.

_____ **8.** There is no way to escape from a rip current.

Reading 2

CURRENTS

In May 1990, a cargo ship was traveling from South Korea to the United States. Soon after it left South Korea, a huge wave swept 21 containers of Nike shoes off the ship and into the waters of the northern Pacific Ocean. Six months later, people on the beaches of northern California, Oregon, Washington, and British Columbia found hundreds of these shoes in the sand. Ocean currents had carried the shoes thousands of kilometers across the Pacific Ocean.

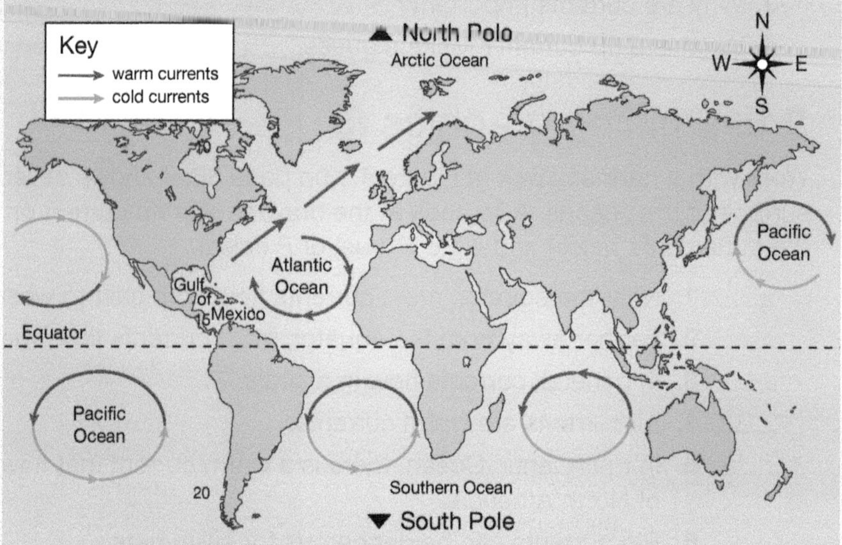

Figure 4.1 Major ocean currents

Surface currents and winds

A current is like a river of warm or cold water that flows through the ocean. The currents in the top layer of the ocean (down to approximately 200 meters) are called surface currents, and they move in different directions. The main cause of surface currents is wind. In general, surface currents in the ocean follow a circular path. They travel west along the equator, turn as they reach a continent, travel east until they reach another area of land, and then go west along the equator again. For example, near the equator, there are winds called tropical trade winds. Trade winds blow from east to west, and they create several currents near the equator. These currents move in a westward direction. Between the poles and the equator, there is another wind system, the westerlies. These winds blow from west to east, and they create currents that move in an eastward direction.

The role of surface currents

Surface currents help spread the heat from the sun around Earth. They move water in big circles. This causes cold water to move to warmer places, and warm water to move to cooler places. It prevents, or stops, warm water near the equator from becoming too hot. It also prevents cold water near the North and South Poles from becoming too cold.

Currents affect the temperature of ocean water and the temperature on land. The moving water of currents heats or cools the air around them. As a result, ocean currents can influence climate. A good example is the Gulf Stream. This huge warm-water current begins in the Gulf of
30 Mexico, flows past the East Coast of North America, and eventually reaches northern Europe. The warm water of the Gulf Stream causes parts of Ireland and England to have warmer temperatures than they would normally have that far north.

Currents and climate change

Since ocean currents influence climate, changes in the way currents
35 move can cause climate change. For example, thousands of years ago, the Gulf Stream stopped flowing. As a result, air temperatures in Europe decreased, and there was a small **ice age** until the Gulf Stream returned. Scientists tell us that similar changes in ocean currents could occur in the future. They continue to study the relationships between
40 oceans and climates. This will help them better understand climate conditions today and the changes that may come in the future.

ice age long period of time in which temperatures gradually decrease and thick sheets of ice cover large areas of land

Rip Currents

Rip currents are small currents that flow away from the shore and out into the ocean. Unlike the Gulf Stream and other huge currents that travel for thousands of kilometers, rip currents flow only a few hundred meters. These currents are small, but they can be extremely dangerous because they travel very fast. A powerful rip current can carry a swimmer too far out into the ocean in less than a minute.

More than 100 people drown in rip currents in the United States every year. This is often what happens: A swimmer feels that he or she is suddenly moving quickly away from the shore. The swimmer gets very nervous and tries to swim back to shore against the powerful current. This is extremely tiring. The swimmer becomes too tired to swim anymore and then drowns.

Rip currents occur at many beaches around the world. Here are some guidelines that will help keep you safe in the ocean:

- Swim only at beaches with lifeguards, and never swim alone.
- If you find yourself in a rip current, stay calm. Do not try to fight the current.
- Swim parallel to (go in the same direction as) the shore until you are out of the rip current.
- If that does not work, try to float and let the current carry you to its end. Then swim back to shore.

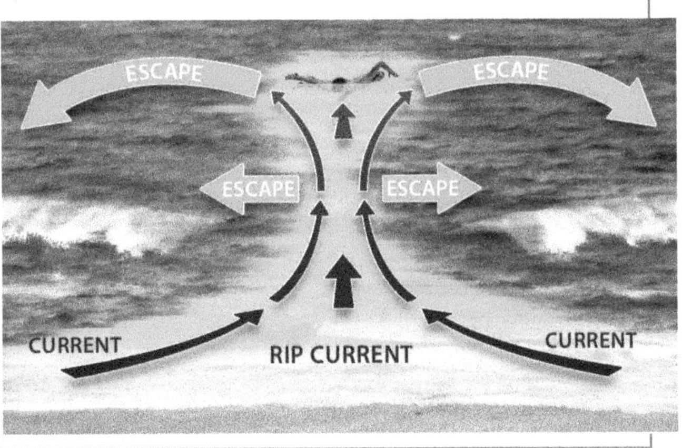

AFTER YOU READ

1 Highlighting A R

Don't forget to highlight. Highlighting is a helpful way to remember important information, such as the main ideas and key terms in a text.

A Highlight the following key words in the reading and the boxed text. Then use the information to write the definition of each term.

currents   Gulf Stream   rip currents   trade winds   westerlies

B Read these questions about the main ideas of the text. Find the answers in the text and highlight them. Use a different color from the one you used in Step A.

1. What causes surface currents?
2. What path do surface currents usually follow?
3. What do surface currents do?
4. How do currents affect water and land?
5. What is the name of a famous warm-water current? Where does it flow?
6. What are rip currents? Why are they dangerous?

2 Labeling a map A R

A Look at this map of ocean currents. Label the compass *N* (north), *S* (south), *E* (east), *W* (west). Locate the following places and label them: equator (*EQ*), North Pole (*NP*), South Pole (*SP*), the westerlies (*WS*), trade winds (*TW*), Gulf Stream (*GS*).

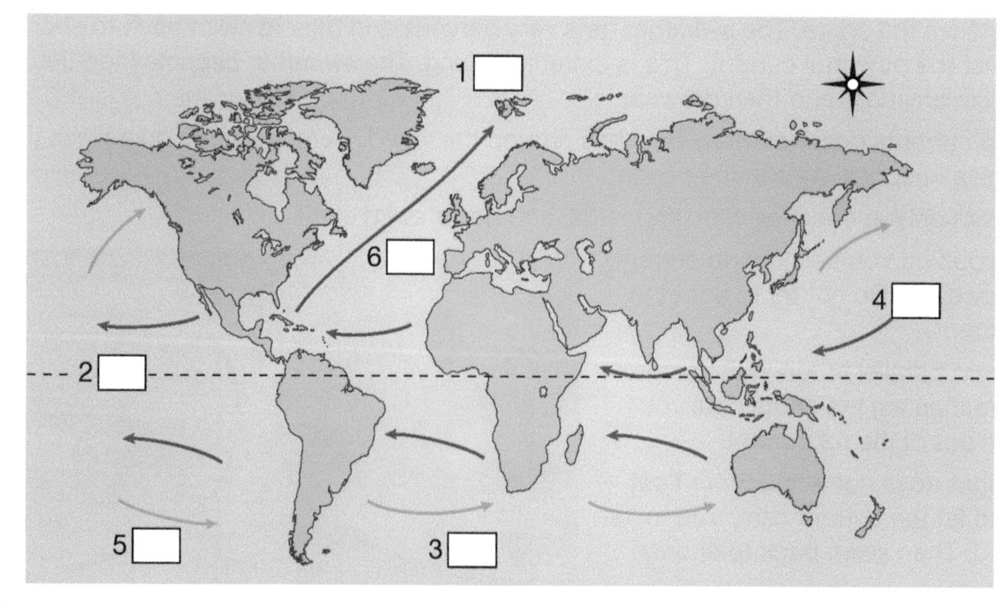

B Look back at the text "Currents" and Figure 4.1 to check your work.

3 Subject-verb agreement V

In sentences beginning with *there is* or *there are*, the *be* verb agrees with the noun after it.

singular form of *be* — singular noun

There **is** a warm **current** off the coast of Alaska.

If a subject has a prepositional phrase after it, the verb agrees with the subject of the sentence, not with the object of the preposition. Many students find it helpful to cross out prepositional phrases in a sentence. This helps them find the subject of the sentence more easily.

prepositional phrase

plural subject — object of the preposition — plural verb

Winds ~~near the equator~~ **blow** from east to west.

A Circle the correct form of each verb in parentheses.

1. There (*is / are*) a cold ocean current near the coast of California.
2. A warm current near the coast of Japan (*flow / flows*) from south to north.
3. Currents by the equator generally (*travel / travels*) in a westward direction.
4. There (*is / are*) winds called tropical trade winds near the equator.
5. Currents in the ocean (*have / has*) an effect on Earth's climates.

B Go back to Step A. Cross out prepositional phrases and underline the subjects.

4 *Too* and *very* V

The words *too* and *very* before an adjective give more detail about a description. *Too* before an adjective usually has a negative meaning. Use *too* to show that something is more than you need or more than you want.

The water is **too** cold. We can't go swimming.

Use *very* to emphasize, or give importance to, an adjective.

The water was **very** cold when we went swimming.

A Go back to the boxed text "Rip Currents." Find uses of *too* and *very* and underline them. Then take turns with a partner and explain the writer's choice for each use.

B Choose the correct word to complete each sentence.

1. The sun was (*too / very*) hot. We couldn't stay out because the heat made us sick.
2. The sun was (*too / very*) hot. We put on sunscreen and enjoyed the warm weather.
3. The waves were (*too / very*) big. It was a good day for surfing.
4. The waves were (*too / very*) big. The lifeguard said it was dangerous to swim.

C Write three sentences about a water activity such as surfing, swimming, or sailing. Use *too* or *very* in each sentence.

PREPARING TO READ

1 Brainstorming Ⓡ

Brainstorming is one way to explore a topic before you read. Often one idea leads to another idea, and this then leads to a different idea. Therefore, it is important to think openly and freely about the topic. When you brainstorm:

- Pose a question and answer it.
- Think of as many ideas as you can.
- Think quickly.
- List all ideas.
- Judge later.
- Organize later.

In a small group, brainstorm the ways that oceans influence people's lives. Try to think of all the ways (both good and bad) that oceans affect us. Set a time limit so that you will think quickly. Choose one person in your group to make a list of everyone's ideas.

2 Organizing ideas Ⓐ

After you brainstorm, it is useful to organize your ideas into categories. This will help you think, talk, and write about the ideas in a logical way.

In your group, organize the ideas that you brainstormed. Use a chart like the one below.

How oceans influence our lives	Ideas from your brainstorming list
Good ways	
Bad ways	

Reading 3

WAVES AND TSUNAMIS

The ocean can be both beautiful and enjoyable. Many people like walking on the beach and watching the water. Others enjoy swimming, surfing, and sailing. However, the ocean is not predictable, and it can be very dangerous. Wind can create big waves that knock people 5 down, sink boats, and damage the shoreline. Giant waves, called tsunamis, can kill people and wash away entire towns. The ocean is truly a place of great beauty and great danger.

Wind and waves

In addition to causing currents, the wind creates waves as it blows across the surface of the ocean. The stronger the wind is, the bigger 10 the waves are. You may remember the beach on a windy day or a calm day. On windy days, big waves crash onto the shore. On a calm, windless day, the surface of the water is flat.

Scientists measure waves from their highest point, the crest, to their lowest point, the trough. Most waves are neither very big nor 15 dangerous. However, some waves, such as those in a violent storm, are large enough to damage ships and hurt people on the shore. Careful swimmers and surfers do not go into the ocean when the waves are too big, and ship captains often change direction to stay away from a storm. The power of the wind and the waves can be deadly.

Tsunamis

20 The biggest, most powerful waves on our planet are called tsunamis. Neither winds nor waves create tsunamis. Tsunamis form when an underwater volcano erupts or an earthquake occurs on the ocean floor. Tsunamis move very quickly across the open ocean. Some tsunamis 25 travel as fast as airplanes – more than 800 kilometers an hour. In the deep ocean, tsunamis do not look like giant waves. In fact, they are usually less than one meter high. However, as they approach land and move into shallow water, they are forced to slow down. This causes the waves to suddenly rise up high in the air and then slam down on the 30 land. In shallow coastal waters, tsunamis can cause waves to rise more than 30 meters high and cause terrible damage to the coast. They kill people and destroy buildings and crops.

In 1960, a huge earthquake occurred off the coast of Chile. Fifteen minutes later, a tsunami hit the Chilean coast. Fifteen hours later, another tsunami hit Hilo, Hawaii. Seven hours after that, another 35 tsunami hit Japan. Thousands of people died because of the tsunamis that occurred after the Chilean earthquake. However, the deadliest series of tsunamis happened on December 26, 2004, when a powerful earthquake occurred in the Indian Ocean. The earthquake caused tsunamis that hit a dozen countries, including Indonesia, India, Sri 40 Lanka, and Thailand. The tsunamis killed more than 250,000 people and destroyed hundreds of towns.

The next time you go to the beach, take a few moments to appreciate both the beauty and the danger of the ocean. Never forget what the power of the waves can do.

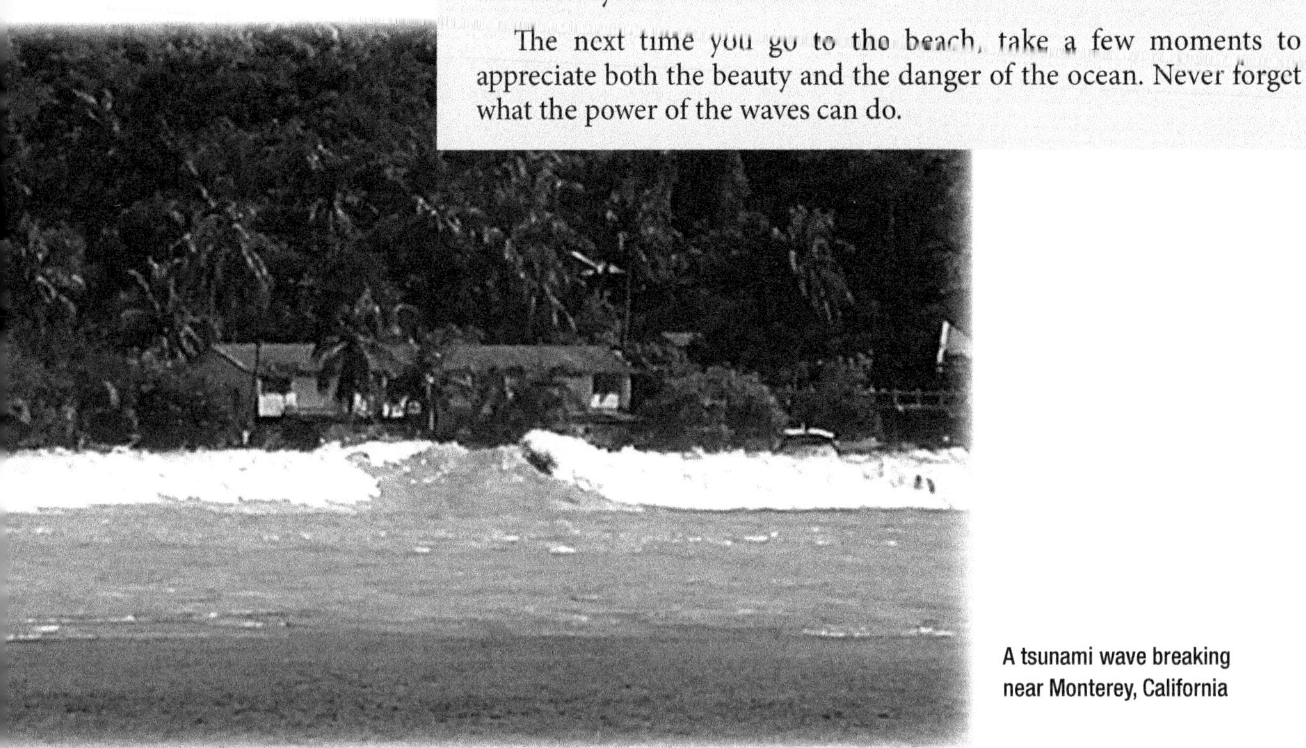

A tsunami wave breaking near Monterey, California

Surviving the 2004 Tsunami

On December 26, 2004, most people had no warning that tsunamis were moving across the Indian Ocean and would soon reach land. As a result, thousands of people died. However, on the Indian island of South Andaman, one group of people, the Jarawa tribe, seemed to know that a tsunami was coming. The Jarawa is one of the oldest tribes on Earth, perhaps 70,000 years old. All 250 members of the tribe escaped safely. They left the coast and went into the jungle before the tsunami arrived. They stayed in the jungle for several days. How did the Jarawa know that the tsunami was coming? The Jarawa people do not like to talk to people outside their tribe. They did not explain. However, some scientists think they have an explanation. The Jawara know the movements of the ocean, earth, and wind, and the behavior of birds. They have a great knowledge of nature. These scientists believe that the ancient tribe used this knowledge to predict the tsunami and survive.

AFTER YOU READ

1 Reading for main ideas and details

The sentences below are main ideas or details from the reading. Write *M* (main idea) or *D* (detail) in the blanks. (Hint: There are three main ideas in the list.)

____ **1.** The wind creates waves as it blows across the surface of the ocean.
____ **2.** On windy days, big waves crash onto the shore.
____ **3.** Storm waves can damage ships and hurt people.
____ **4.** Tsunamis are the biggest waves on our planet, and they can cause terrible damage.
____ **5.** Tsunamis can travel 800 kilometers per hour.
____ **6.** The deadliest series of tsunamis happened on December 26, 2004.
____ **7.** The ocean is both beautiful and dangerous.
____ **8.** Tsunamis rise high in the air as they approach land.

2 Adjective suffixes

Many nouns and verbs can be changed into adjectives by adding suffixes such as *-ful*, *-able*, and *-ous*. Some words are both nouns and verbs. Look at these examples:

Noun and / or verb	Adjective
truth (n)	truthful
prevent (v)	preventable
love (n, v)	lovable
mountain (n)	mountainous

A Read the list of nouns and verbs. Find the adjective forms in the reading and circle them. Then write the adjective form of each word in the blank.

	Adjective form	
1. beauty (n)	______________	(Par. 1)
2. enjoy (v)	______________	(Par. 1)
3. predict (v)	______________	(Par. 1)
4. danger (n)	______________	(Par. 1)
5. care (n, v)	______________	(Par. 3)
6. power (n)	______________	(Par. 4)

B Complete the sentences with adjectives from Step A.

1. Swimming at the beach on a sunny day can be ________.
2. It is too ________ to be on the beach when a tsunami reaches land.
3. Rocket ships need ________ engines to travel into space.
4. A ________ swimmer never goes swimming alone.
5. Earthquakes are not ________, but you can prepare for them.
6. Last night the sunset over the ocean was very ________.

C Work with a partner. On a separate piece of paper, write four or five sentences that describe a water feature such as a tsunami, an ocean, or a river. Use the adjectives in Step A.

3 Parallel structure W

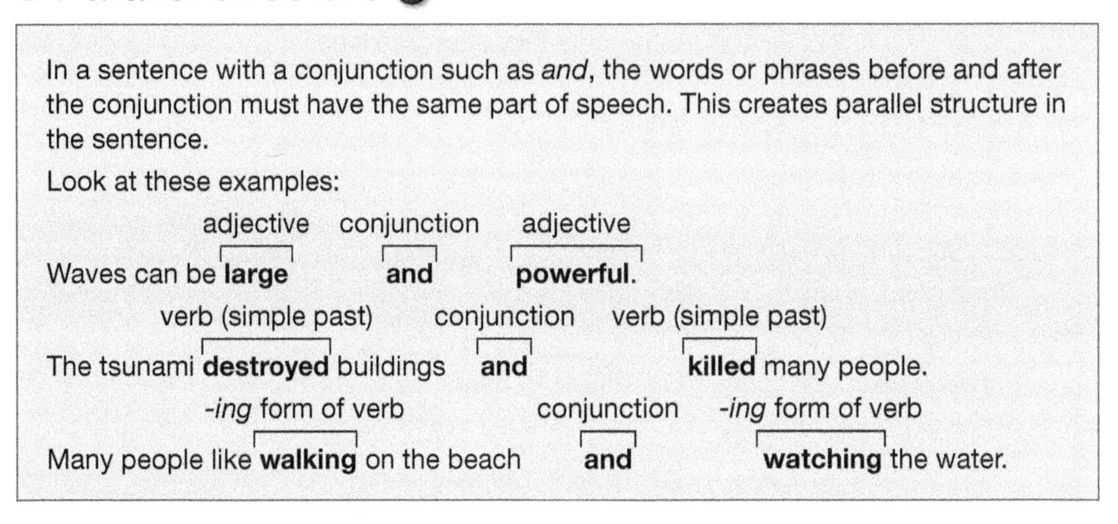

In a sentence with a conjunction such as *and*, the words or phrases before and after the conjunction must have the same part of speech. This creates parallel structure in the sentence.

Look at these examples:

Waves can be **large** (adjective) **and** (conjunction) **powerful** (adjective).

The tsunami **destroyed** (verb (simple past)) buildings **and** (conjunction) **killed** (verb (simple past)) many people.

Many people like **walking** (*-ing* form of verb) on the beach **and** (conjunction) **watching** (*-ing* form of verb) the water.

A Reread paragraph 1 of "Waves and Tsunamis." Find five examples of parallel structure and underline them. Then compare answers with a partner.

B Find the error in parallel structure in each sentence. Use the strategies in the box above. Write the correct sentences on a separate piece of paper. Look back at the text to check your answers.

1. The ocean can be beautiful and enjoyment.
2. Many people like walking on the beach and to watch the water.
3. The wind can create big waves that knock people down, sink boats, and damaged the shoreline.
4. The power of the windy and the waves can be deadly.
5. The tsunamis killed more than 250,000 people and destroying hundreds of towns.

4 *Both . . . and* and *neither . . . nor*

Both . . . and and **neither . . . nor** are two-part conjunctions that require parallel structure. In other words, the parts of speech that follow each conjunction must be the same. Look at these examples:

	conjunction	adjective	conjunction	adjective
The Great Lakes are	**both**	**deep**	**and**	**wide**.

conjunction	noun	conjunction	noun	
Neither	**lakes**	**nor**	**rivers**	are as salty as oceans.

A Look back at the text "Waves and Tsunamis." With a partner, find sentences with *both . . . and* and *neither . . . nor* and mark them. How many did you find? Which words have the same form in each sentence?

B Complete the sentences with a word from the box. Be sure to use parallel structure.

dangerous	preventable	Southern	volcanoes	waves

1. Wind creates both currents and __________.
2. The Pacific Ocean is violent. It has both earthquakes and __________.
3. Neither the Arctic Ocean nor the __________ Ocean is as large as the Atlantic Ocean.
4. Earthquakes are neither predictable nor __________.
5. The ocean is both beautiful and __________.

5 Reviewing paragraph structure

Read the incomplete paragraph below. Complete the topic sentence and add a concluding sentence.

Duke Kahanamoku ____

__________________ . He was born in 1890 in Honolulu, Hawaii, and he spent his whole life near the ocean. Kahanamoku enjoyed swimming and canoeing, but he was most famous for his skill in surfing. In fact, many people consider him the father of surfing. He traveled frequently and introduced surfing to people all over the world. He won five Olympic medals in swimming. When Kahanamoku was born, very few people surfed. By the time he died, surfing was a sport that millions of people enjoyed.

____

____ .

Duke Kahanamoku

Chapter 4 Academic Vocabulary Review

The following words appear in the readings in Chapter 4. They all come from the Academic Word List, a list of words that researchers have discovered occur frequently in many different types of academic texts. For a complete list of all the Academic Word List words in this chapter and in all the readings in this book, see the Appendix on page 206.

affect	factors	parallel	similar
appreciate	guidelines	removed	varies
approach	jobs	series	widespread

Complete the sentences with words from the list.

1. Rip currents and surface currents are ________ in that they both consist of moving water.
2. The temperature of ocean water ________ depending on location. Water near the poles is colder than water near the equator.
3. Pollution is a ________ problem. Human actions have dirtied the water, land, and air in many places on our planet.
4. After a large natural disaster, people are sometimes ________ from the area. They cannot return until the area is safe.
5. Many people like to walk ________ to the ocean along the beach. That way, they can enjoy the wonderful scenery without getting wet.
6. Experts predict that the number of ________ in health care and education will increase in the future.
7. Students decide on a field of study based on many ________, including personal interest and career potential.
8. Do not ________ a monk seal on the beach. Monk seals are endangered, and it is illegal to get too close to them.
9. Garbage in the ocean can ________ ocean animals and many people such as fishermen.
10. The more people learn about the interconnectedness of life on Earth, the more they ________ the need to protect our planet.

Practicing Academic Writing

In this unit, you learned about water on our planet. Based on everything you read and discussed in class, you will write a paragraph about this topic.

A Water Feature

You will write one academic paragraph about any water feature on Earth. For example, you may write about the location and size, history, or special features of an ocean, lake, or glacier. Use information from this unit and your own ideas. Find additional information in the library or on the Internet.

PREPARING TO WRITE

Choosing a topic, exploring ideas, and making a simple outline

There are several steps to completing a writing task. The first is preparing, and preparing is also divided into steps.

- First, make sure you understand the assignment.
- Next, choose a topic and explore it.
- Third, make a simple outline to organize your ideas.

Each step is important. For example, exploring helps you decide if you are really interested in the topic and have enough information to complete the assignment. Stating and summing up your ideas in a simple outline helps you understand the ideas and the basic relationships between them.

A Work with a partner or in a small group. Read the writing prompt again. Answer the following questions.

1. How much are you supposed to write?

a. a sentence
b. a paragraph
c. two paragraphs
d. an essay

2. What are you supposed to write about?

a. an ocean and a lake
b. a volcano
c. a glacier
d. a water feature

3. Where should you find extra information to help you write about the topic?
   a. in the textbook
   b. from your own knowledge
   c. from your classmates
   d. from the library and Internet

Note: If you have any questions about the assignment, ask your teacher now.

B Now that you understand the assignment, it is time to choose a topic.

1. Look at the photographs of water features to get ideas. Then, with your partner or group, make a list of familiar water features. Include the water features discussed in this unit as well as others, such as waterfalls, seas, gulfs, bays, streams, ponds, straits, canals, and channels.

2. On your own, choose two water features that interest you. On a piece of paper, brainstorm the first water feature for five minutes. Then, on a separate piece of paper, brainstorm the second water feature that you chose. When you brainstorm, ask yourself lots of who, what, where, why, and how questions such as these:
   - Where is the water feature?
   - How old is it?
   - How did it form? Did nature or humans create it?
   - Who discovered it?
   - Is it famous? Is it important? Why or why not?
   - How big is it?
   - Has it changed over the years? In what ways?
   - Is it dangerous? Why?
   - Are people concerned about this water feature? Why?

   Remember to list every idea.

3. With your partner or group, discuss the two water features you chose.
4. Choose the water feature that most interests you and find more information about it. You can interview other people or do research in the library or on the Internet. Take notes on what you find.
5. Review your brainstorming list and notes. Then choose three important ideas that you want to include in your paragraph. Think of at least one detail or example to support each idea.
6. Fill in the simple outline below. This will help you organize your ideas before you begin to write.

   Topic Sentence: ______________________________

   Main Idea #1: ______________________________

   Main Idea #2: ______________________________

   Main Idea #3: ______________________________

NOW WRITE

Write the first draft of the paragraph.

- Write a topic sentence that states the main idea of the paragraph.
- Write five to seven supporting sentences based on the outline you created.
- End the paragraph with a concluding sentence that restates the main idea. (Be sure to make the concluding sentence a little different from the topic sentence.)
- Remember to use correct paragraph form and structure.
- Try to include some of the vocabulary you learned in this chapter.
- Make sure the subject and verb in each sentence agree with each other.
- Give your paragraph a title.

AFTER YOU WRITE

After you write your paragraph, look for ways to improve it. One way to do this is to make sure you wrote on topic.

Writing on topic

After you write a paragraph once (first draft), read it again to make sure that all the sentences are on topic. Identify any irrelevant sentences – that is, sentences that do not support the topic – and cross them out. This will make the paragraph stronger and more focused. Then rewrite the paragraph (second draft).

A Read the following paragraph. It has two irrelevant sentences. One is crossed out. Find the other sentence and cross it out.

One of the most important currents on Earth is the Antarctic Circumpolar Current. This current flows from west to east in the Southern Ocean. Strong westerly winds blow the current around the continent of Antarctica through the waters of the Atlantic, Pacific, and Indian Oceans. ~~These three oceans also have many other currents.~~ The Antarctic Circumpolar Current is the largest ocean current in the world, and it moves more water around the globe than any other current. It also keeps warm ocean water away from Antarctica. Ocean water is warm near the equator. That's why the ice there does not melt.

B Read the following paragraph. Find two irrelevant sentences. Cross them out.

Like the Gulf Stream Current, the Humboldt Current has a strong effect on the climate of the land it flows past. This cold-water current travels south along the west coast of South America, from northern Peru to the southern end of Chile. Air temperatures in Chile are cooler than we expect because of this ocean current. The California Current makes the climate of the Hawaiian Islands cooler than we might expect, too. The Humboldt Current also affects climate in another way: It makes areas of northern Chile, southern Peru, and Ecuador extremely dry. Peru and Ecuador are countries in South America.

C Reread your paragraph and cross out any irrelevant sentences.

D Exchange paragraphs with a partner and read each other's work. Then discuss the following questions about both paragraphs:

- Which idea in your partner's paragraph do you think is the most interesting?
- Does your partner's paragraph have a topic sentence that states the main idea?
- Are all the supporting sentences on topic? Are there any irrelevant sentences?
- Are major and minor supporting details included?
- Does the paragraph have a concluding sentence that restates the main idea?

E Think about any changes to your paragraph that might improve it. Then write a second draft of the paragraph.

Unit 3

The Air Around Us

In this unit, you will look at the air that surrounds our planet and at Earth's weather and climates. In Chapter 5, you will examine what Earth's atmosphere is made of, why it is important, and how it is structured. You will also learn about clouds. In Chapter 6, you will focus on weather conditions in the atmosphere. You will look at different climates around the world and at several types of storms.

Contents

In Unit 3, you will read and write about the following topics.

Chapter 5 Earth's Atmosphere	Chapter 6 Weather and Climate
Reading 1 The Composition of the Atmosphere	**Reading 1** Climates Around the World
Reading 2 The Structure of the Atmosphere	**Reading 2** Storms
Reading 3 Clouds	**Reading 3** Hurricanes

Skills

In Unit 3, you will practice the following skills.

Reading Skills

Previewing key terms
Building background knowledge about the topic
Thinking about the topic
Previewing key parts of a text
Examining graphics
Previewing art
Applying what you have read
Increasing reading speed
Reading for main ideas

Writing Skills

Reviewing paragraph structure
Transition words
Writing about height
Writing an observation report
Introducing examples

Vocabulary Skills

Guessing meaning from context
Describing parts
Playing with words
Colons, *such as*, and lists
Words from Latin and Greek
When clauses
Defining key words
Using a dictionary
Using *this* / *that* / *these* / *those* to connect ideas
Synonyms
Prepositions of location

Academic Success Skills

Examining test questions
Taking notes with a chart
Using symbols and abbreviations
Understanding averages
Using a Venn diagram to organize ideas from a text
Examining statistics
Thinking critically about the topic

Learning Outcomes

Write an academic paragraph about the climate in a place you know

Previewing the Unit

> **Previewing** means looking at one thing before another. It is a good idea to preview your reading assignments. Read the contents page of every new unit. Think about the topics of the chapters. You will get a general idea of how the unit is organized and what it is going to be about.

Read the contents page for Unit 3 on page 102 and do the following activities.

Chapter 5: Earth's Atmosphere

A Go outside or open a window and then do the following. Work with a partner. Take notes and share ideas.

1. Take a deep breath of air. The air is part of our planet's atmosphere. Can you see the atmosphere? Can you smell it or feel it? Is the atmosphere the same everywhere?
2. Next, look up into the air. Do you see birds? What do you see? Make a list of everything that you see in the atmosphere.
3. There are some things in the air that you cannot see, such as rockets that carry astronauts far beyond Earth. Add some of these things to your list.
4. Do you see clouds? What can clouds tell us about the weather?

B Discuss the results of your activities in Step A in a small group or as a class.

Chapter 6: Weather and Climate

A People often describe air temperature, wind, rain, or snow to talk about weather. In a small group, describe the weather in your hometown. Compare climates with your group.

B Storms are an important part of weather conditions. Work in a small group to complete these activities.

1. Read these names of storms. Check (✓) the storms that are common in your area.

   ____ rainstorms
   ____ snowstorms
   ____ thunderstorms
   ____ hurricanes
   ____ tornadoes

2. Discuss big storms that you remember. Then answer these questions:
   - Which type of storm is the most common where you live now?
   - Which type of storm is most common on our planet?
   - Which type of storm is the most deadly?

Chapter 5
Earth's Atmosphere

PREPARING TO READ

1 Previewing key terms R V

A The first reading in this chapter is "The Composition of the Atmosphere." What does *composition* mean? Look up the word in a dictionary if you do not know.

B Read this definition of *atmosphere*. Then work with a partner and answer this question: What is the atmosphere? Use your own words to explain. Do not repeat the definition below.

> The atmosphere is a blanket of gases that covers every part of our planet Earth. It is the air around us.

C Now explain the title of the reading. What does "composition of the atmosphere" mean? Exchange ideas with your partner.

2 Building background knowledge about the topic R

A The atmosphere is composed of 12 different gases. What names of gases do you know? Make a list. Are any of these gases part of the atmosphere?

B People often use chemical symbols to represent gases. For example, the symbol for argon is Ar. Work with a partner and complete the following activities.

1. Look at this list of six gases in the atmosphere. Match the gases with the symbols.

____	**1.** hydrogen	**a.** O_3
____	**2.** oxygen	**b.** CO_2
____	**3.** nitrogen	**c.** He
____	**4.** ozone	**d.** H_2
____	**5.** helium	**e.** N_2
____	**6.** carbon dioxide	**f.** O_2

2. How much do you know about these gases? Try to answer these questions.

1. Which two gases make up most of Earth's atmosphere?
2. Which gas makes balloons float?
3. Which gas protects life on our planet from harmful light from the sun?
4. Which gas makes soda bubbly?
5. Which gas is very flammable (catches fire easily)?

Reading 1

THE COMPOSITION OF THE ATMOSPHERE

The atmosphere is a blanket of gases that covers every part of our planet Earth. It is the air around us. You cannot see air, but you can feel it when the wind blows. 5 Every time you take a breath, air goes into your lungs.

What goes into your lungs when you breathe? Our air is composed of a mixture of 12 gases: nitrogen, oxygen, 10 argon, carbon dioxide, water vapor, neon, helium, methane, krypton, hydrogen, ozone, and xenon. The two main gases are nitrogen and oxygen. The atmosphere consists of 78 percent nitrogen and 15 21 percent oxygen. The other gases make up only a small percentage of the atmosphere, but they are very important.

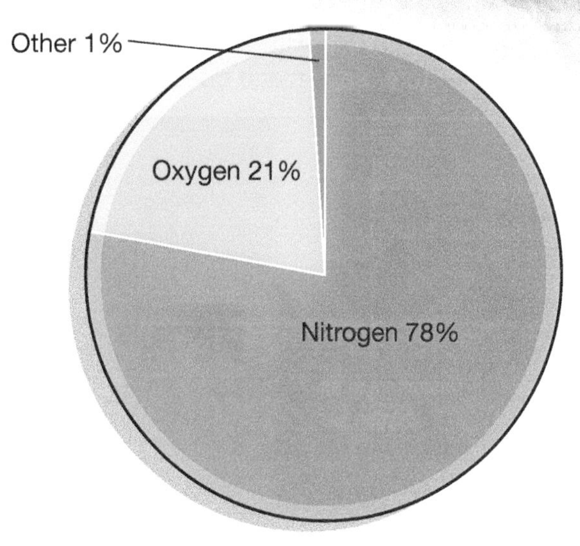

Figure 5.1 The gases in Earth's atmosphere

meteor a piece of rock or metal from outer space that makes a bright light in the sky as it falls into Earth's atmosphere

Humans need the atmosphere for many reasons. First, our bodies need the oxygen in the air to keep us alive. Second, humans need 20 plants, and plants need the nitrogen in the air to grow. In addition, the atmosphere acts like a shield around Earth. It protects us from objects that fall from space, such as **meteors**. The ozone in the atmosphere also protects us. It blocks harmful rays from the sun; without ozone, the rays would burn us. Finally, some gases help control temperatures on 25 Earth. Carbon dioxide, for example, prevents the air from becoming too cold. The special combination of gases in the atmosphere allows life on Earth to exist.

Oxygen

Oxygen is one of the main gases in Earth's atmosphere. Its chemical symbol is O_2 (as an element its symbol is O). Oxygen has no color, odor, or taste. It makes up 21 percent of Earth's atmosphere, 90 percent of the water in our oceans, and nearly 50 percent of Earth's crust. It also makes up about 60 percent of the human body. Almost all living things need oxygen.

Oxygen also has other important uses. For example, factories use oxygen to produce steel, plastics, and textiles such as cotton and silk. Space rockets use liquid oxygen for fuel. Doctors may give nearly pure oxygen to patients with breathing problems. In addition, airplanes and spacecraft use a mix of oxygen with other gases that helps people breathe better at high altitudes.

Today, oxygen even provides recreation. "Oxygen bars" appeared in Japan in the late 1990s; now they are in airports, health clubs, and many other places. By 2012, some bars were even selling "flavors" of oxygen, such as lavender and peppermint. Customers believe that the oxygen improves energy and lowers stress. For about US $1 a minute, they breathe air with more oxygen than the usual 21 percent in the atmosphere. Some stores now sell cans of oxygen for people to enjoy anywhere. Does oxygen help with altitude sickness, jet lag, and headaches? The can manufacturer says that it does, but scientific studies have not proven these benefits.

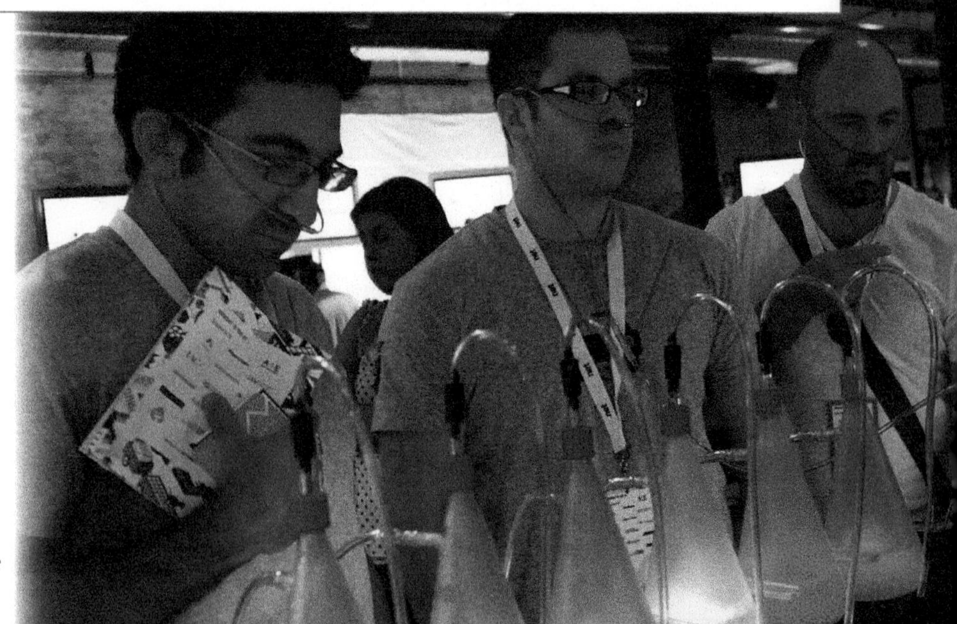

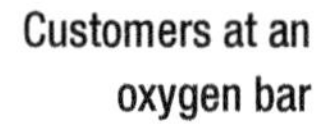
Customers at an oxygen bar

AFTER YOU READ

1 Examining test questions (A) (R)

Always examine the language of test questions. Test questions may express ideas differently from the way they are expressed in a reading.

A Read the questions. Find the answers in the reading "The Composition of the Atmosphere."

1. How many gases combine to create the atmosphere?
2. Which are the two most common gases?
3. Which gas do plants need to grow?
4. Which gas helps keep us safe from the sun?
5. Which gas prevents the air from becoming too cold?
6. What makes life on Earth possible?

B Work with a partner. Underline the words and phrases in the reading that answer the questions in step A. Discuss the differences between the language in the questions and the language in the reading.

2 Guessing meaning from context (V)

The context of a new word, that is, the words and phrases that surround it, can often help you guess the meaning.

Find the words on the left in the reading "The Composition of the Atmosphere." Use the context of each word to guess the meaning. Then match the words and definitions.

____ **1.** blanket
____ **2.** lungs
____ **3.** shield
____ **4.** meteor
____ **5.** rays

a. beams of light from the sun
b. something that covers something else to protect it from harm or damage
c. the parts of the body that help a person breathe
d. a layer that covers something
e. a bright light from a piece of rock or metal when it falls from space into Earth's atmosphere

3 Describing parts Ⓥ

There are many ways to say what a thing is made of. Learn the following words. The phrases are used to describe content or to say that a thing has parts.

composed of (*X is composed of A, B, and C.*)

consists of (*X consists of A, B, and C.*)

make up / made up of (*A, B, and C make up X. / X is made up of A, B, and C.*)

A Go back to the reading. Find three phrases that say that a thing has parts. Circle them.

B Complete these sentences about the composition of the atmosphere. Use information from the text, Figure 5.1 on page 105, and the boxed text on page 106. Fill in each blank with one word.

1. Earth's ________ consists ________ 78 percent ________ and ________ percent oxygen.
2. Ten gases ________ ________ less than one ________ of the atmosphere.
3. The ________ is ________ of 12 ________.
4. The atmosphere is ________ ________ several gases that are essential for human life.
5. Oxygen ________ ________ almost ________ percent of Earth's crust.
6. The atmosphere ________ of less than ________ ________ carbon dioxide.

4 Reviewing paragraph structure Ⓦ Ⓡ

Remember that many paragraphs in academic writing have the following structure:

- The **topic sentence** is usually the first sentence. It tells readers the main idea of the paragraph, with no details.
- **Supporting sentences** follow the topic sentence. They provide specific details and examples to show that the topic sentence is true. They also explain the topic sentence more fully. They are put in logical order.
- The **concluding sentence** is the last sentence. It repeats the main idea of the paragraph, but it is not a copy of the topic sentence.

Go back to the reading. Find the paragraph with an "academic" structure. That is, the paragraph must have a topic sentence, supporting sentences, and a conclusion. Check (✓) it. Compare your answers with a partner's.

5 Transition words (W)

It is important to organize ideas in your writing. This helps a reader understand your ideas. After you write the topic sentence, consider how to organize the supporting sentences.

Transition words show the connection between ideas. Look at these transitions and their meanings:

First,	the most important idea; the first that occurs in time
Second,	the next most important idea; the next that occurs in time
In addition,	a point or idea that is added to a previous one
Also,	a point or idea that is added to a previous one
Finally,	the last idea

A Read the sentences and find the transition words. Then put the ideas in order. Number the sentences 1–8. Do not look at the text.

_____ **a.** In addition, the atmosphere acts like a shield around Earth.
_____ **b.** The special combination of gases in the atmosphere allows life on Earth to exist.
_____ **c.** Second, humans need plants, and plants need the nitrogen in the air to grow.
_____ **d.** Humans need the atmosphere for many reasons.
_____ **e.** First, our bodies need the oxygen in the air to keep us alive.
_____ **f.** Finally, some gases help control temperatures on Earth.
_____ **g.** It protects us from objects that fall from space, such as meteors.
_____ **h.** The ozone in the atmosphere also protects us. It blocks harmful rays from the sun . . .

B Write a paragraph with transitions. The topic is the atmosphere. Go back to the text and Figure 5.1 for information. Do not copy from the text. Follow these steps.

1. Begin your paragraph with a topic sentence that states your main idea, without details. This makes a claim about the atmosphere.
2. Give one or more details to prove or illustrate your main idea, and use complete sentences. This is your support.
3. Use transition words to order your information.
4. End with a concluding sentence that restates the main idea.
5. Compare your paragraph with a partner's. Check that the information is ordered correctly.

PREPARING TO READ

1 Thinking about the topic Ⓡ

Where does the atmosphere start? Read the following quotation and see if you change your answer.

> Look at your feet. You are standing in the sky. When we think of the sky, we tend to look up, but the sky actually begins at the earth. – Diane Ackerman

2 Previewing key parts of a text Ⓡ

> Remember to preview key parts of a text before you read. Previewing will help you understand the main ideas.

Preview the title, the introductory paragraph (par. 1), and the headings in the reading on pages 111–112. Then answer the following questions with a partner.

1. What is the text about?
2. How many layers does the atmosphere have?

3 Examining graphics Ⓡ

Look at Figure 5.2 on page 111. Then answer the following questions in a small group.

1. What is the name of the bottom layer of the atmosphere?
2. What is the name of the top layer of the atmosphere?
3. Which layer is the largest?
4. In which layers do airplanes fly?
5. In which layer of the atmosphere do satellites orbit Earth?
6. Which layer contains living things?
7. In which layer can we see meteors?

Reading 2

THE STRUCTURE OF THE ATMOSPHERE

The atmosphere around Earth extends far above the surface of the planet. No clear boundary, or line, marks the end of the atmosphere. The air just gets thinner and thinner. In other words, the upper atmosphere has less 5 oxygen, and it eventually blends into space. Scientists divide the atmosphere into five layers: the troposphere, the stratosphere, the mesosphere, the thermosphere, and the exosphere.

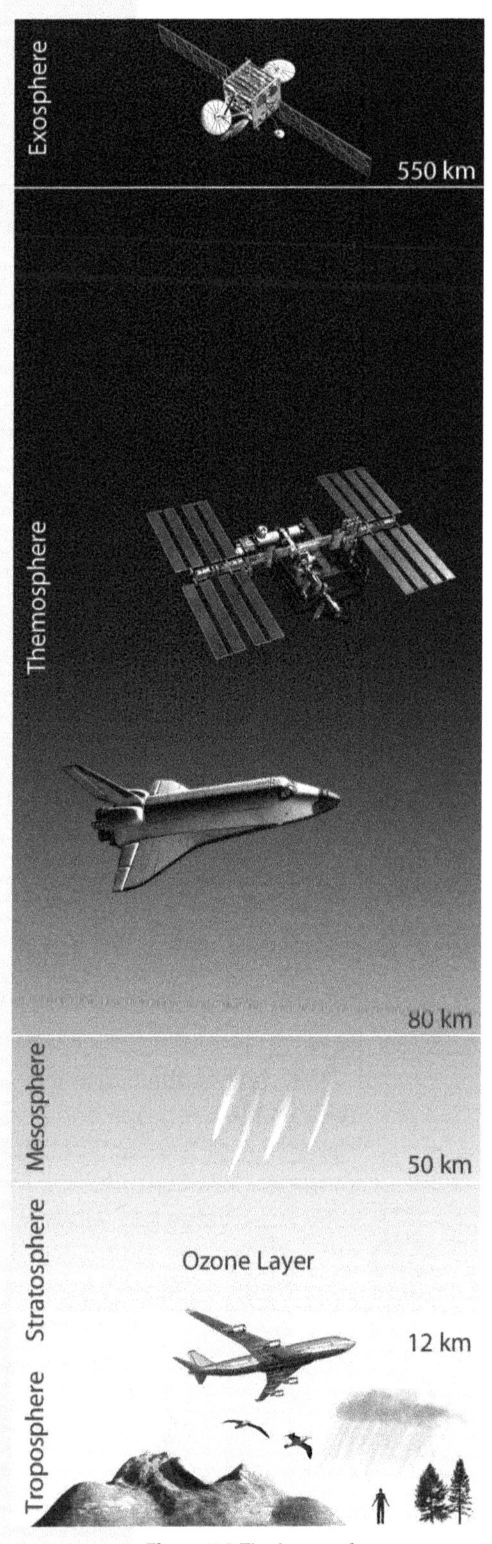

Figure 5.2 The layers of the atmosphere

The troposphere

The troposphere is the first layer of the atmosphere. It 10 extends from Earth's surface to an average of 12 kilometers above the surface. There is a lot of movement and activity in this layer. It contains all the familiar parts of our world: the oceans, the mountains, the clouds, and all living things. Most of the water in the atmosphere is located here, so 15 weather conditions, such as rain, snow, and thunder, occur in the troposphere. In addition, the air in the troposphere is always moving. The movement creates turbulence, that is, a very unstable flow of air.

The stratosphere

20 The second layer of the atmosphere is the stratosphere. It starts at 12 kilometers and ends at about 50 kilometers above Earth. There is no wind or weather in the stratosphere, and there are few clouds. The air is very stable (still) and clear. As a result, pilots sometimes fly in the stratosphere, above 25 the troposphere, to enjoy a smooth ride. The ozone layer is in the stratosphere. The ozone layer is very important for life on Earth. It absorbs dangerous ultraviolet radiation, or invisible rays of energy that come from the sun. Without the ozone layer, humans and animals would probably die 30 from the sun's radiation.

The mesosphere

About 50 kilometers above Earth, the stratosphere ends, and the mesosphere begins. The mesosphere is the third layer of the atmosphere. The mesosphere extends from 50 kilometers to 80 kilometers above the surface of Earth. The 35 air in this layer becomes very thin, and the temperature drops to as low as -93°C. In fact, the mesosphere is the coldest layer of the atmosphere. Most people will never go to the mesosphere, but sometimes we can see events

International Space Station a structure in space that many countries built together to support space research

satellite (made by humans) an object that goes into space to collect information or to become part of a communications system

happening there. Every day millions of meteoroids enter our 40 atmosphere from outer space, and they burn up in the mesosphere. Sometimes we can see their trails as shooting stars in the night sky.

The thermosphere

The fourth layer of the atmosphere is the thermosphere. The thermosphere is located approximately 80 kilometers above Earth's surface. In this layer, the temperature starts to increase again. 45 In fact, the thermosphere is the hottest layer of the atmosphere. Astronauts in their spacecraft orbit Earth in the thermosphere. The **International Space Station** also orbits Earth in the thermosphere. Otherwise, humans do not travel to this layer.

The exosphere

Finally, approximately 550 kilometers above Earth, we reach the 50 exosphere. The exosphere is the last layer of the atmosphere. At the edge of the exosphere, the air becomes extremely thin. Is there any sign of human activity this high up? Yes. Thousands of **satellites** from several countries orbit Earth in this layer.

The next time you look up at the sky, remember this: Each layer of 55 the atmosphere is important. Together the layers produce breathable air, water, and protection from harmful things in space. As far as we know, Earth may be the only planet with atmospheric layers that can support developed life.

The International Space Station

In 1998, construction began on the International Space Station (ISS). Space agencies from Canada, Europe, Japan, Russia, and the United States built the Station. Its mission is to "enable long-term exploration of space and provide benefits to people on Earth."

We can see the ISS in the night sky. It orbits Earth 16 times a day – the same distance as going to the moon and back every day. It is about the size of a football field, or 110 meters by 49 meters. The Station has a modular, or piece-by-piece, design. It has a command module, storage module, labs, gym, living quarters, two bathrooms, and a large bay window. Spacecraft from Earth deliver equipment and supplies.

In November 2010, the ISS marked the 10th anniversary of continuous human occupation. More than 200 people from more than 15 countries and regions, such as the United States, Russia, Canada, Japan, Europe, South Korea, Malaysia, and South Africa, have steadily come to the ISS. Crew members usually stay three to six months.

Crew members operate and maintain the ISS. They conduct astronomy and meteorology experiments and research how living in space affects human health. The ISS also inspires future scientists, engineers, artists, and explorers from around the world.

Astronauts Acaba, Padalka, and Revin traveled 52,906,428 miles and orbited Earth 2,000 times.

AFTER YOU READ

1 Taking notes with a chart (A) (R)

Put your notes in chart form. Charts can show the information in a way that is easy to remember. A chart can help you review information quickly for a test.

A Look at the chart. It shows the beginning of a student's notes on the reading "The Structure of the Atmosphere." Notice that the chart has four columns. The columns include the names, height, and special features of each layer of the atmosphere.

Complete the chart with information from the text. In the last column, try to write two pieces of information for each layer of Earth's atmosphere.

Layer	Name	Height	Special Features
1			• •
2		from 12 km to ____ km	• •
3	mesophere		• •
4			• •
5		from ____ km to ?	• satellites in this layer •

B Compare your chart with a partner's.

2 Playing with words (V)

Play games with words. This will help you learn and remember them. Play with words' similarities and differences and have fun. This is a good way to develop your knowledge and understanding of new and familiar words.

This game is called Odd One Out. Work with a partner. Look at the words in each row. Choose the word that does not belong and explain why.

1. spacecraft	bird	satellite	plane
2. spacecraft	bird	satellite	tree
3. spacecraft	bird	cloud	tree
4. spacecraft	plane	astronaut	space station
5. rain	snow	cloud	thunder
6. atmosphere	exosphere	mesosphere	stratosphere

3 Colons, *such as*, and lists Ⓥ

Use a colon (:) after a complete sentence to introduce a list.

Air is composed of a mixture of 12 different gases: **nitrogen, oxygen, argon, carbon dioxide, water vapor, neon, helium, methane, krypton, hydrogen, ozone, and xenon.**

Use *such as* to introduce a list that is not complete. That is, *such as* introduces items that are examples from a larger list. Notice these things about *such as*: It is not followed by a colon, it does not begin a sentence, and a comma often comes before it.

Air is composed of a mixture of many different gases, **such as nitrogen, oxygen, argon, neon, and helium.**

Notes:

In lists of more than two items, there is a comma after each item.

The word *and* comes before the last word in a list.

A comma is used before *such as* to introduce lists that restrict or limit meaning.

Remember that a list is not a complete sentence – the first word is not capitalized.

A Look back at paragraphs 1 and 2 of the reading "The Structure of the Atmosphere." Find two lists that follow a colon and one list that follows *such as*. Underline them.

B Find the error in each sentence below. Discuss why the sentence is incorrect with a partner. Then, on your own, rewrite it correctly.

1. There are four main types of wet weather: rain, snow, hail, sleet.
   There are four main types of wet weather: rain, snow, hail, and sleet.
2. There are several types of wet weather such as: rain, snow, and hail.
   ______
3. There are four main types of wet weather: rain, snow, and hail, and sleet.
   ______
4. There are several types of wet weather. Such as rain, snow, and hail.
   ______
5. There are several types of wet weather, rain, snow, and hail.
   ______
6. There are several types of wet weather such as, rain and, snow and, hail.
   ______
7. There are many types of wet weather: Rain, snow, hail, and sleet.
   ______

C Write two sentences about Earth's atmosphere. Use a list with a colon in one sentence. Use a list with *such as* in the other sentence. Compare sentences with your partner.

4 Writing about height W R V

Writing about height means writing about how high something is. Height is often included in a description of a thing. When you write about height or altitude in science you may also describe the location and range of something. Here are some structures you can use to express how high something is, including its location and range:

- A is located X above B.
  The exosphere **is located** 550 kilometers **above** Earth's surface.
- A extends from X to Y above B.
  The mesosphere **extends from** 50 kilometers **to** 80 kilometers **above** Earth.
- A starts / begins at X and ends at Y above B.
  The thermosphere **starts at** approximately 80 kilometers **and ends at** 550 kilometers **above** Earth.

A Find the expressions of height in the reading. Underline them. Compare your answers with a partner's.

B Complete these sentences about the layers of the atmosphere. Include the structures in the box. Include words and numbers. Check your answers in the text and Figure 5.2.

1. The ________ extends ________ our planet's surface ________ an average of 12 kilometers above the surface.
2. The ________ is located 30 kilometers above the stratosphere.
3. The mesosphere ________ at 50 kilometers and ________ at 80 kilometers above Earth.
4. The thermosphere ________ located ________ kilometers ________ Earth.
5. Satellites ____________ hundreds of kilometers above the planet in the ________.

C Now write four sentences of your own about the atmosphere. Use an expression of height in each sentence.

1. ______________________________

____

2. ______________________________

____

3. ______________________________

____

4. ______________________________

____

PREPARING TO READ

1 Previewing art Ⓡ

A Look at the photos in the reading on pages 117–118 then look at the drawings of these clouds.

a. ____________________

b. ____________________

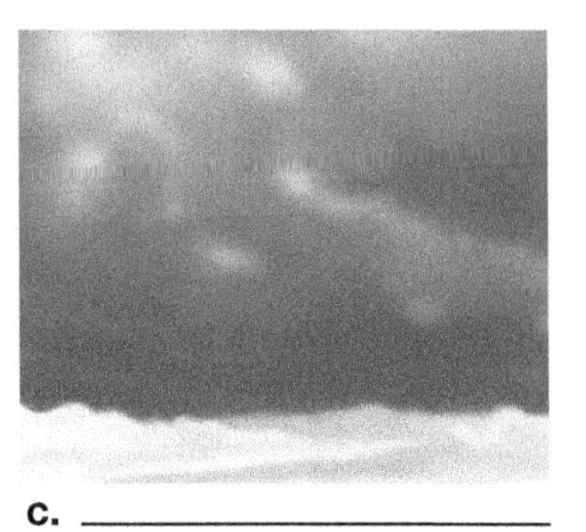

c. ____________________

B Answer the questions with a partner.

1. What are the names of the clouds? Write each name on the blank in step A.
2. Do you know the names of other clouds? Name the clouds you know.
3. Describe the weather when you see the clouds in Step A. Use the words in the box to help you.

bad weather	cold	good weather	rain	sunny	warm
blue sky	fog	ice	snow	stormy	wet

2 Building background knowledge about the topic Ⓡ

A Study the pictures below. They show the three forms (states) of matter that water can take.

liquid

solid

gas

B Describe the states of matter that water can take in Earth's atmosphere. Write *gas*, *liquid*, or *solid* in the blanks.

1. Water in ice is in the form of a __________.
2. Water in raindrops is in the form of a __________.
3. Water that is neither a liquid nor a solid is a __________. (It is called water vapor.)

Reading 3

CLOUDS

Clouds are a familiar sight in the troposphere. We see them in the sky almost every day. Do you ever wonder what they are made of? Clouds are composed of billions of tiny water droplets or ice crystals. How are they formed? Consider this special feature of water: It can change from a gas (vapor) to a liquid (water) to a solid (ice) and back again. Clouds form when warm water vapor in the air rises in the atmosphere. As the warm water vapor moves higher into cooler air, the air in the vapor cools. It then becomes tiny drops of water or ice. These droplets of water or ice join together in the sky and form clouds. Clouds come in many different shapes and sizes, but there are three main types: cumulus, cirrus, and stratus.

Cumulus clouds

Cumulus clouds are fluffy, white clouds. They look like balls of cotton in the sky. Children often draw cumulus clouds in their pictures. When you see cumulus clouds, the weather is generally good, and the sky is blue. Cumulus clouds are low-level clouds. They are usually close to the ground, about 460 to 915 meters above Earth's surface. A cumulus cloud forms when sunshine warms water vapor in the air. The warm water vapor rises and cools as it moves higher. The cool water vapor then changes into water droplets, and the droplets join together and form a cumulus cloud.

Cumulus clouds

Cirrus clouds

Cirrus clouds are thin, wispy white clouds. They look like the tail of a horse or a curl of hair. When you see cirrus clouds in the sky, it usually means that stormy weather is on its way. Cirrus clouds are located very high in the sky. They are about 5 to 12 kilometers above the ground in the troposphere. It is very cold high in the troposphere, so cirrus clouds are made up of tiny ice crystals, not water droplets. A cirrus cloud forms when cold air moves under an area of warm air. The cold air pushes the warm air higher in the troposphere. The warm air cools, changes into ice crystals, and a cirrus cloud forms.

Cirrus clouds

Stratus clouds

Stratus clouds

Stratus clouds look like gray, shapeless blankets that cover most of the sky. They are usually only about 0.8 kilometer thick, but they can be almost 1,000 kilometers wide. When you see stratus clouds, you might soon see rain. Stratus clouds are low-level clouds that are located quite close to the ground. In fact, sometimes they lie on the ground or the ocean, and then they are called fog. A stratus cloud forms when warm, wet air moves slowly over an area of cooler air. The warm air rises, cools, and changes into water droplets. The droplets join together and create a stratus cloud.

There are many other types of clouds. Each one forms in a different way, and each one can tell you something about the coming weather. If you pay careful attention to the types of clouds in the sky, you can predict the weather fairly accurately.

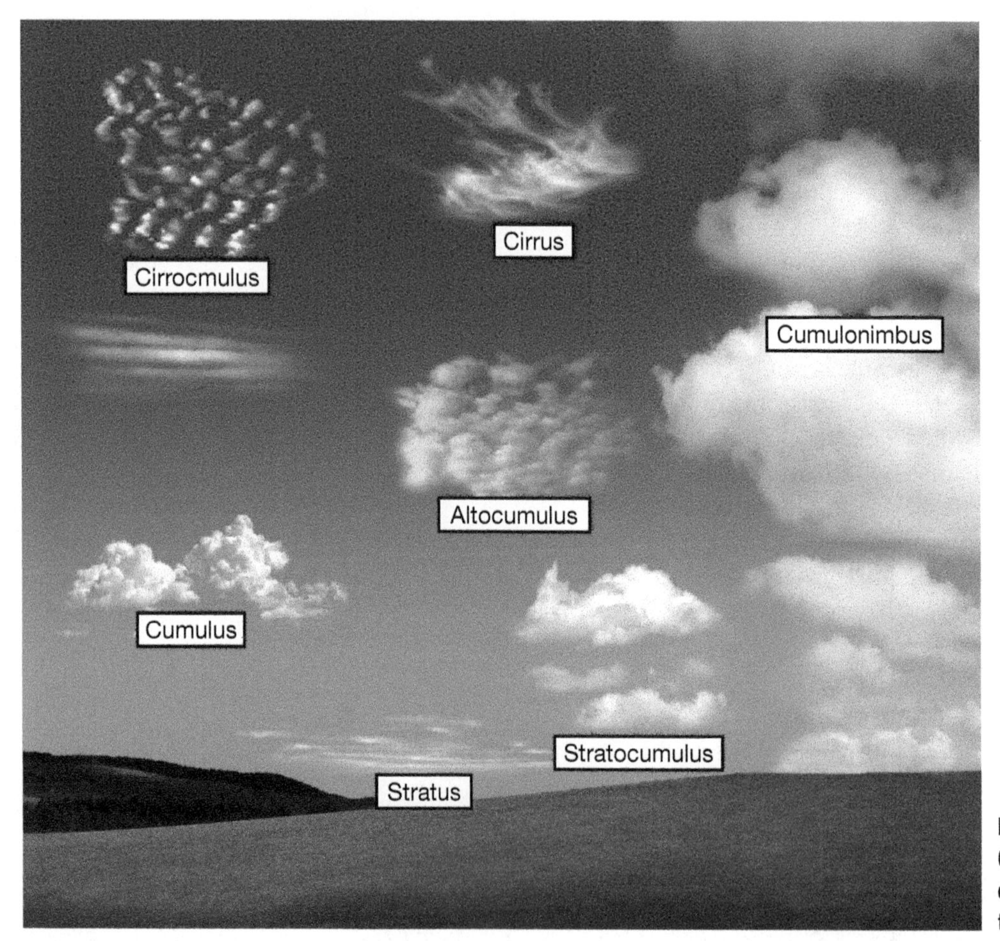

Figure 5.3 Common types of clouds in the troposphere

The Latin origin of cloud names

Scientists often use Greek or Latin words to name plants, animals, and other things in nature. Cloud names come from the Latin language. *Cumulus* is the Latin word for a large heap or lump. *Cirrus* means "curl" in Latin, and *stratus* refers to a blanket or layer. *Cumulus*, *cirrus*, and *stratus* can also combine with other Latin words. For example, *nimbus* means "rain" or "storm," so *cumulonimbus* clouds are big heaps of rain clouds. Another example is *alto*. Although *alto* means "high," scientists use it as a prefix to describe midlevel clouds. For instance, altostratus clouds are layers of clouds that are neither very high nor very low in the sky. Finally, *asperatus* is the Latin word for "rough." Scientists may have discovered a new type of cloud. This cloud looks like choppy ocean waves that cover the sky. They named these new formations asperatus clouds. After you learn the Latin words, you will be able to understand the name of almost every cloud in the sky.

Asperatus clouds

AFTER YOU READ

1 Taking notes with a chart A R

The chart below shows the beginning of a student's notes on the reading "Clouds." Fill in the missing information. In the last column, draw a picture of each type of cloud.

Cloud name	Description	Picture
Cumulus		
	• thin, wispy, white • looks like a thin curl of hair • found 5–12 km above ground	

2 Using symbols and abbreviations A W

It is important to write quickly when you take notes. As in texting messages, you do not use complete words or sentences. Try to abbreviate (shorten) words and use symbols to represent them. Look at these examples of common abbreviations and symbols. You can also create your own to make your notes especially meaningful to you.

Abbreviation		**Symbol**	
m	*meters*	&	*and*
@	*at*	≈	*approximately*
b/c	*because*	→	*causes or leads to*

A Work with a partner. Guess the meanings of the symbols and abbreviations below. Then add three more to the list.

1. info ________ **4.** + ________ **7.** ____ ________

2. ex ________ **5.** ↑ ________ **8.** ____ ________

3. km ________ **6.** = ________ **9.** ____ ________

B Share your answers with your classmates.

C Below are some notes on a paragraph in the reading "Clouds." Work with a partner. Take turns using the notes to make sentences about cumulus clouds.

Cumulus clouds
- fluffy & white
- usually = *good weather*
- low level (≈ 460–915 m above ground)
- sun warms water vapor → water vapor rises → water vapor cools → water droplets = cumulus clouds

D Complete these notes on the paragraph about cirrus clouds.

Cirrus clouds
- thin, wispy, white
- usually = ________________ soon
- high level (________________)
- b/c cold @ high level, made of ice, not ________________
- cold air moves under warm air → warm air ________________ → warm air cools → ________________ = ________________

E Now write your own notes on the paragraph about stratus clouds on a separate piece of paper.

3 Words from Latin and Greek V

Remember that many scientific words come from Latin or Greek.

A Read the names below. These are two more types of clouds. Guess what these clouds look like. Use the information from the boxed text on page 119 to describe them.

- nimbostratus
- cirrostratus

B Draw a picture of each cloud in Step A. Compare your drawings with a partner.

C Look at the following list of word parts from Greek and Latin and their meanings.

astro-	star	*-graph*	write, written	*photo-*	light
bio-	life	*-logy*	study	*-scope*	observe, see
geo-	earth, rock	*-meter*	measure	*tele-*	far, distant

D Work with a partner and guess the meanings of the following words. Then check your answers in a dictionary.

1. telescope
2. geology
3. astrometry
4. biometrics
5. biology
6. astrophotography
7. telephoto
8. photometer

E What other words can you make in English with the word parts in the box? Make a list and share it with the class.

4 *When* clauses 

Sentences in English often have more than one clause. A clause is a group of words that has a subject and a verb. In the examples below, each sentence has a *when* clause and a main clause.

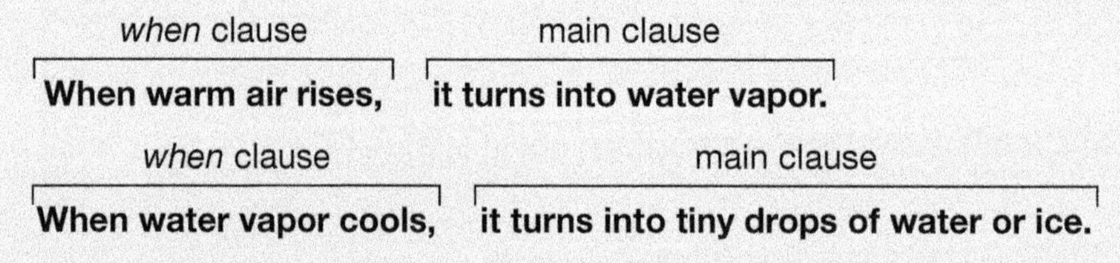

The *when* clause and the main clause can also be in the reverse order. In that case, there is no comma between the two clauses.

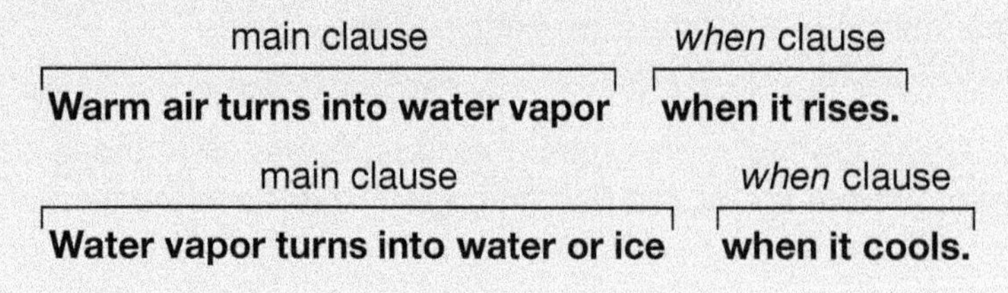

A Find six sentences in the reading with a *when* clause. Underline the sentences. Do commas separate the *when* clause from the main clause? In how many sentences?

B Complete the sentences with a *when* clause or a main clause. Use information from the reading and the boxed text. Use commas correctly. Compare your sentences with a partner's.

1. When water vapor rises ____.
2. It is probably not going to rain ____.
3. When children draw clouds ____.
4. We call it fog ____.
5. When you see cumulonimbus clouds ______________________________________.

5 Writing an observation report W

Students in science classes often have to write observation reports. In these reports, they record what they see and examine that information carefully.

Write an observation report about clouds and weather. Follow the instructions below. Work with a partner on parts 1 and 2.

1. Go outside or look out the window and look up at the sky. What is the weather like today? What kinds of clouds do you see? Use many details to describe the weather and clouds.
2. Do your observations of the clouds and the weather agree with the information in the text "Clouds"? Explain why or why not.
3. Write a report about your observations. You can use the model below for help, but remember that it is just a brief example. Write as many details as you can about the clouds and the weather. Include a picture of the different types of clouds you saw.

Name: ____________________

Date: ____________________

Observation Report

Today there are ____________________ clouds in the sky. They are ____________________ clouds. According to the reading "Clouds," when you see ____________________ clouds, the weather is usually ____________________. This is true / not true about the weather today. Today's weather is, in fact, ______________________________.

Chapter 5 Academic Vocabulary Review

The following words appear in the readings in Chapter 5. They all come from the Academic Word List, a list of words that researchers have discovered occur frequently in many different types of academic texts. For a complete list of all the Academic Word List words in this chapter and in all the readings in this book, see the Appendix on page 206.

conduct	equipment	maintain	symbol
construction	finally	stress	unstable
enable	invisible	structure	vehicles

Complete the following sentences with words from the list.

1. Headphones, transmitters, and antennas are part of the __________ for special radios on the International Space Station.
2. Some companies are building electric __________ to try to reduce the air pollution from cars and trucks that use gasoline.
3. Hikers should watch out for __________ rocks. Loose rocks can make people slip and fall.
4. Many dentists earn a high salary, but they may also experience a high degree of __________ because of their jobs. In fact, problems with patients cause many dentists to lose sleep at night.
5. Five layers of air compose the __________ of the atmosphere.
6. For many people, a white dove is a __________ of peace.
7. The International Space Station took more than 12 years to build, but it was __________ finished in 2011.
8. Sometimes air pollution is __________, but other times people can actually see black smoke from factories and cars in the air.
9. The __________ budget of most space agencies is very large. It costs a lot of money to build spacecraft.
10. Improvements in technology __________ astronauts to video chat with students in schools on Earth.

Developing Writing Skills

In this section, you will practice taking notes and using notes to write a paragraph. You will also use what you learn here to complete the writing assignment at the end of this unit.

Using Your Own Words When You Write

Writing in college courses means showing that you understand what you read. Using your own words to write about ideas is a skill. The key is taking notes. Here are some guidelines:

- Read the text once to understand the main ideas. Read the text again and take notes.
- Do not take notes on every detail. Focus on main ideas, major supporting details, and a few specific examples and facts.
- Use symbols and abbreviations to put ideas in your own words.
- Close your book and practice writing out ideas in your notes.
- Explain the main ideas to someone else. Use your notes, not the book.
- Use your notes, not the book, to complete a writing assignment.

A Use the strategies in the box to complete the activities.

1. Read the paragraph.

A *meteoroid* is a piece of rock or metal that travels through outer space. Meteoroids can be as tiny as dust or as large as 10 meters. However, most are small, like a pebble. Meteoroids move very quickly. The fastest meteoroids travel about 42 kilometers per second. A *meteor* is a piece of rock or metal from outer space that makes a bright light in the sky as it falls into Earth's atmosphere and burns up. Millions of meteoroids enter Earth's atmosphere every day, and their meteors look like "shooting stars" in the night sky. A *meteorite* is a meteoroid that lands on the ground. More than 100 meteorites hit Earth each year.

2. Read the paragraph again and take notes. Then write sentences from your notes. Use your own words. Remember to use symbols and abbreviations.

3. Work with a partner. Look at the notes below that a student took on the same reading. Some are correct and some are not. Check (✓) the two notes that are NOT correct and rewrite them.

____ **a.** Meteoroid = rock or metal from space

____ **b.** Fastest meteoroids go 42 km/hr

____ **c.** Shooting stars = meteors = meteroids burning up in atmos.

____ **d.** Meteorites >100 fall to Earth each yr

4. Compare the corrected notes to your own notes. Do they match?

B Remember that your assignment is to write a paragraph from notes. The topic of your paragraph is auroras. An aurora is a natural sight in the atmosphere.

1. Look at the photograph and read the notes about auroras.

Auroras

- auroras = beautiful lights in sky
- usually green-yellow (can be red, blue, violet)
- diff. shapes & sizes
- occur in thermosphere
- 100–300+ km above Earth
- particles from sun + atmospheric gases above N. and S. poles ➔ auroras
- ex. of famous auroras:
    1. aurora borealis = Northern Lights best seen from Alaska, E. Canada, Iceland (Sept.–Oct. & Mar.)
    2. aurora australis = Southern Lights best seen from Antarctica

Aurora borealis

2. Discuss the meanings of the symbols and abbreviations in the notes with a partner. Use a dictionary to look up the new words. Work alone and write the notes out into complete sentences. Then take turns expressing the ideas in your own words with your partner.

3. Now write a paragraph about auroras. Remember to:

- include a topic sentence, at least five supporting sentences, and a concluding sentence.
- order ideas logically. Use transitions to guide the reader through your paragraph.
- use *when* clauses and expressions of height where appropriate.

4. Exchange paragraphs with a partner. Check for a clear topic sentence with support, transitions, *when* clauses, correct descriptions of height, and a conclusion. Check that ideas are stated correctly and match the notes shown in item 1.

5. Consider ways to improve your paragraph. Write a second draft. Correct any errors in spelling and grammar.

Chapter 6
Weather and Climate

PREPARING TO READ

Thinking about the topic Ⓡ

Work with a partner or in a small group and do the following activities.

A Look at the title of the chapter. How can you describe weather and climate? Brainstorm words that name and describe weather, such as *wind*, *rain*, *hot*, *cold*, *cloudy*, and *sunny*. Make a list. Think of at least 15 words.

1. Describe the weather today. Use words that you brainstormed.
2. Look at the photograph and describe the weather.

B Think about weather that is similar in different places.

1. How is the weather in the photograph similar to the weather where you live? How is the weather different?
2. Read this list of places. Do any have similar weather? Divide the places into three groups. List them in the chart below.

Alaska (U.S.A.)
Gobi Desert, China
Hawaii (U.S.A.)
Nord, Greenland
Northern Canada
Puerto Rico
Sahara Desert, Africa
Rub' al-Khali, Saudi Arabia
Thailand

Group 1	Group 2	Group 3

3. Discuss your chart with your classmates. Explain your reasons for your groupings.

Reading 1

CLIMATES AROUND THE WORLD

What is the weather like today? Is it hot or cold? Is it sunny, rainy, or snowy? Weather is the atmospheric conditions at a particular time. Climate is different from weather. Climate means the average weather conditions of an area over a long period of time, at least 30 years. An area's climate includes its average temperature and its average amount of precipitation. Climate determines the kinds of plants and animals that live in an area. For example, tropical rain forests grow in hot, wet climates, and polar bears live in cold climates.

Scientists divide Earth into climate zones, or areas, according to temperature and precipitation. Some of the main climate zones are tropical, dry, mild, and polar. Tropical climates are located near the equator. They have warm temperatures all year and a lot of rain. Hawaii, Puerto Rico, and Thailand all have tropical climates. Dry climates have very little precipitation, so they do not have a lot of plant life. The Gobi Desert in China and the Sahara Desert in Africa are good examples of dry climates.

Mild climates have neither very hot nor very cold temperatures. It rains, but not as heavily as in tropical climates. For example, San Francisco and London are cities with mild climates. Polar climates are 20 the coldest areas on Earth. Even during the warmest months, average temperatures are below 10° Celsius (50° Fahrenheit). Polar climates are also very dry, with less than 38 centimeters of precipitation each year. The Arctic (including parts of Alaska, Canada, Greenland, and other areas around the North Pole) has a polar climate.

25 Scientists tell us that Earth's climate is changing in important ways. For example, air and ocean temperatures have been rising in recent years. This change is called **global warming**. Scientists believe that global warming is causing other climate changes on our planet, such as an increase in heat waves and more powerful storms all over the world.

global warming the increase over time in the temperature of Earth's atmosphere

Cherrapunji: A Place of Extremes

Does it rain very much where you live? Do you get tired of wearing a raincoat and carrying an umbrella? If so, be glad you don't live in Cherrapunji, India. Some people think Cherrapunji may be the wettest place on Earth. In fact, it is one of the rainiest places on our planet, with an average rainfall of more than 1,143 centimeters per year. It rains in Cherrapunji all year; in fact it is the only place in India that receives rain year round. However, most of the rain falls during six months of the year. Cherrapunji actually has a shortage of water during the other six months. How is this possible?

Cherrapunji is a mountain town in the northeastern state of Meghalaya. For six months of the year, winds called monsoons blow from the southwest, and they bring extremely heavy rain. The other half of the year, the monsoons blow from the northeast; they bring some rain, but not very much. In the past, the area around Cherrapunji was mostly forest. The trees protected the soil and kept it in place. Therefore, the ground could hold water from the rainy season for use throughout the year. However, in recent years the town grew, and people cut down large numbers of trees; they wanted space for new homes and businesses. Now that the trees are gone, most of the rainwater runs down into the valley. It no longer goes into the ground. That's why one of the rainiest places in the world must bring in water from other areas during the dry season.

Cherrapunji, India

AFTER YOU READ

1 Applying what you have read R A

Apply what you have read to new subject matter. This can show how well you understood the text.

Study the chart below. Look at each place, with its average temperature and precipitation. Then predict the climate. Write *tropical*, *dry*, *mild*, or *polar*. Go back to the reading if necessary.

Place	Average annual temperature	Average annual precipitation	Climate
1. Manila, Philippines	27°C / 81°F	206 cm / 81.1 in	
2. Inuvik, Canada	-9.5°C / 15°F	27 cm / 10.6 in	
3. Namib Desert, Namibia	16°C / 61°F	5 cm / 2 in	
4. Yakutsk, Russia	-10°C / 14°F	20 cm / 7.9 in	
5. Monrovia, Liberia	26°C / 79°F	513 cm / 202 in	
6. Santiago, Chile	14°C / 57°F	38 cm / 15 in	

2 Defining key words V

In academic texts, look for the definitions of key words in the margins as well as in the text.

Look at the list of key words from "Climates Around the World." Find the words in the reading. Then match the words and definitions.

1. _____ precipitation
2. _____ equator
3. _____ climate zone
4. _____ Celsius
5. _____ global warming

a. a scale for measuring temperature
b. an increase in the temperature of Earth's atmosphere
c. rain, snow, sleet, and hail
d. an imaginary line around the middle of Earth
e. an area of land that has similar average temperatures and amounts of precipitation in every part

3 Understanding averages Ⓐ

An average is a general number that represents a level, amount, or degree of something that is usual for a group or class of people or things. Scientific texts often use averages for things that can be measured in numbers, such as temperature or precipitation. To calculate an average, add the number of each member of the group and get a total, then divide the total by the number of members in the group. For example, the average temperature of 20°C, 10°C, and 30°C is **20**°C. Add 20° + 10° + 30° = 60° and divide 60° ÷ 3 = **20**°.

A Look at Amman, Jordan's average monthly temperatures.

Average Monthly Temperatures in Amman

January	8°C	July	25°C
February	9°C	August	26°C
March	12°C	September	24°C
April	16°C	October	21°C
May	21°C	November	15°C
June	24°C	December	10°C

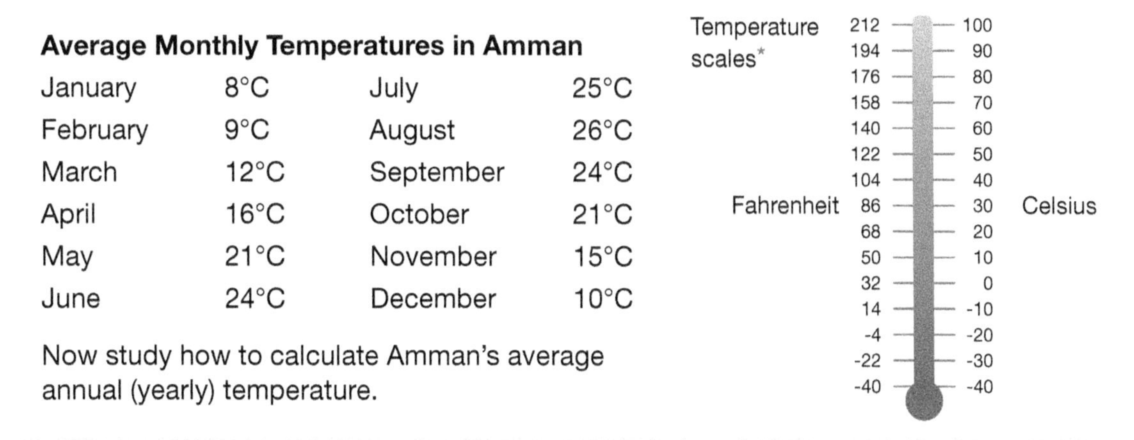

Now study how to calculate Amman's average annual (yearly) temperature.

Average annual temperature = sum of monthly temperatures ÷ 12 months:

Monthly temperatures: 8° + 9° + 12° + 16° + 21° + 24° + 25° + 26° + 24° + 21° + 15° + 10°
Total of monthly temperatures = 211°
Total of monthly temperatures, divided by 12 months = 211° ÷ 12 = 17.6°

The average annual temperature in Amman, Jordan, is 17.6° C (63.7° F).

B Calculate the average annual temperature in Buenos Aires, Argentina, from the information below. Write the answer on a separate piece of paper in a complete sentence.

Average Monthly Temperatures in Buenos Aires

January	23°C	May	13°C	September	13°C
February	22°C	June	10°C	October	16°C
March	20°C	July	10°C	November	19°C
April	16°C	August	11°C	December	22°C

C Calculate the average summer temperatures in Amman (June, July, August) and Buenos Aires (December, January, February). Write the answers in a compound sentence on your piece of paper from step B.

---

* *You can find formulas for converting one temperature scale to another on page 205.*

4 Introducing examples Ⓦ

Good writing shows and tells. It includes details to prove a point. Include specific examples to show the reader what you mean. Specific examples also make your writing stronger and more interesting. Look at these two phrases that introduce examples:

For example introduces a sentence:

For example, rain forests grow in hot, wet climates.

Such as introduces a noun or a list of nouns within a sentence:

Rain forests have many types of animals, **such as** gorillas, parrots, and crocodiles.

Notice that there is a comma after *for example* and a comma before *such as*.

A Go back to the reading. Mark uses of *for example* and *such as* to introduce examples.

B Read this text about global warming.

Our planet is getting warmer. Some temperature change is natural. However, temperatures on Earth increased much faster than expected in recent years.

Global warming is causing some troubling climate changes. Unfortunately, people around the world are adding to the problem. They are using more and more energy for heat, electricity, and transportation. Burning fossil fuels produces most of this energy, and this increases global warming. To slow global warming and help prevent further damage to the planet, people need to make some changes in their daily lives.

C The sentences and phrases below give examples that explain ideas in the text in Step B. Match the examples to the ideas. Then rewrite the text from Step B on a separate piece of paper with the new examples below.

1. For example, they could recycle more things, walk more, and drive less.
2. For example, some areas are having more heat waves, others are getting heavier rain, and polar areas are getting warmer.
3. such as oil, gas, and coal
4. For example, over the 100 years of the twentieth century, temperatures increased by 0.5°C. However, before that time, it took 400 years for temperatures to increase by the same amount.

PREPARING TO READ

Previewing key parts of a text Ⓡ

A Preview the title, the headings, and the photographs on pages 134–135 to get a general idea of the reading. Then read the first two sentences of each paragraph to get more specific ideas.

Answer the following questions.

1. What is the text about? Check the correct answer.

_____ two kinds of storms that cause a lot of damage

_____ ideas about two kinds of dangerous storms

_____ descriptions and statistics for two types of storms

2. What two types of storms does the text discuss? Write the names of the storms next to the pictures below.

a. ______________________________

b. ______________________________

B Write *T* (true) or *F* (false) for each statement.

_____ **1.** There are only a few thunderstorms every year.

_____ **2.** Most thunderstorms do not last a long time.

_____ **3.** Thunderstorms can cause tornadoes.

_____ **4.** Tornadoes are most common in Europe.

Reading 2

STORMS

Thunderstorms

At this moment, almost 2,000 thunderstorms are occurring around the world. Thunderstorms are different from rain showers. Rain showers do not create thunder and lightning, but thunderstorms do. **Lightning** is electricity that moves between clouds, or between a cloud and the ground. Lightning heats the air around it. The hot air expands. Then it quickly contracts as it cools down. This movement of air makes the sound called thunder. Thunder and lightning happen at almost exactly the same time. However, light moves faster than sound, so people see lightning before they hear thunder.

lightning a flash of bright light in the sky, caused by electricity during a thunderstorm

Approximately 90 percent of thunderstorms are small and last no longer than 30 minutes. These short thunderstorms bring cool rain on a hot, humid day. However, 10 percent are powerful storms that can produce hail, strong winds, heavy rain, and tornadoes. Severe thunderstorms can continue for hours and even days, and they can cause a lot of damage. Lightning strikes occur over 40 million times every year, and every year, lightning causes over 7,000 forest fires, hundreds of injuries, and more than 90 deaths in the United States. Hail is another destructive and costly problem. Hail consists of balls of ice called hailstones. Hailstones can be as small as a pea or as large as a grapefruit. Hail causes almost $1 billion of damage each year. However, the most dangerous part of severe thunderstorms is the heavy rain. Heavy rain can cause flooding, and flooding can be more dangerous to people and property than lightning or hail.

Tornadoes

Severe thunderstorms can also cause tornadoes. Tornadoes, or "twisters," are tall, spinning, funnel-shaped clouds that touch the ground. They usually last 15 minutes or less, but they move quickly and can cause a great deal of damage in just a few seconds. The air inside a tornado spins very fast, more than 480 kilometers per hour, as it moves quickly upward. Tornadoes can lift trees, cars, people, animals, and even houses into the air.

Tornadoes occur throughout the world. These dangerous storms are most common in the United States. More than 1,000 tornadoes occur each year in the United States. Most of them occur in "Tornado Alley." Tornado Alley is an area in the central part of the country. The states of Texas and Oklahoma experience the most tornadoes in this area. On average, more than 100 tornadoes strike Texas each year. That is approximately the same number of tornadoes that strike the whole country of Canada in one year.

Tornado "seasons" arve difficult to predict. For example, in 2012, an unusual number of tornadoes occurred in the United States in January. However, May, normally the most active month, was very quiet. Today, scientists still have many questions about severe thunderstorms and tornadoes. They continue to do research in order to better understand, predict, and prepare people for these destructive storms.

Green Skies

Often the sky is blue. Sometimes it is gray. But have you ever seen a green sky? Yes, green skies are real, but they are rare.

Many people believe that a green sky is a sign of a tornado. Scientists cannot find evidence to support this belief, but they do say that there is evidence of green thunderstorms. Usually people see green skies during a severe thunderstorm. Although scientists are not sure why the sky turns green, many think it is because thunderstorms have a lot of water. Water is naturally blue, and sunlight is often red, especially at sunset. When the large amount of blue water mixes with the red sunlight, the sky can look green. In fact, green skies are most common in the late afternoon and evening.

Whatever the reason the sky turns green, it usually means that a dangerous thunderstorm is close by. Often, hail and tornados occur with the thunderstorm. Therefore, if you see green skies, you should move to a safe place as quickly as you can.

AFTER YOU READ

1 Using a Venn diagram to organize ideas from a text 

Science and other types of academic texts often discuss classes and types of things. They present related information, or information that is common to a class, and information that is specific to a member of the class. A Venn diagram organizes information from related texts and can show the following:

- in one circle, information that is true only for topic A
- in the other circle, information that is true only for topic B
- in the overlapping part of the circles, information that is true for both topics (A+ B)

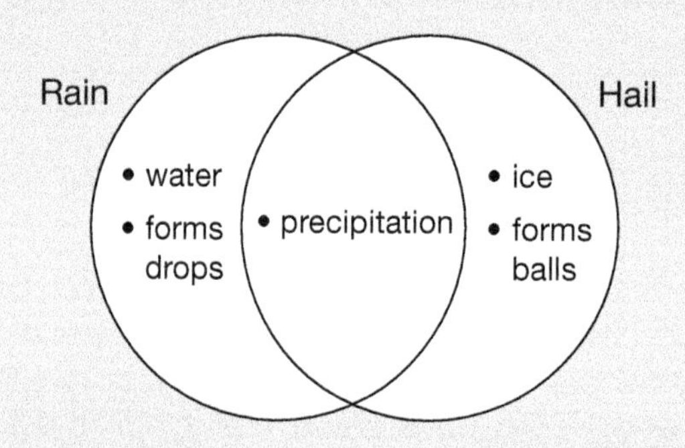

A Look at the information from "Storms." Sort facts that are true only for thunderstorms, only for tornadoes, or for both types of storms. Write *TH* for thunderstorms, *TO* for tornadoes, or *B* for both storms.

____ are fast moving
____ can cause a lot of damage
____ are tall, spinning clouds
____ happen more than 1,000 times a day
____ can lift houses into the air
____ can produce lightning and hail
____ happen throughout the world
____ can cause dangerous flooding

B Complete the diagram. Write the facts from Step A in the correct places.

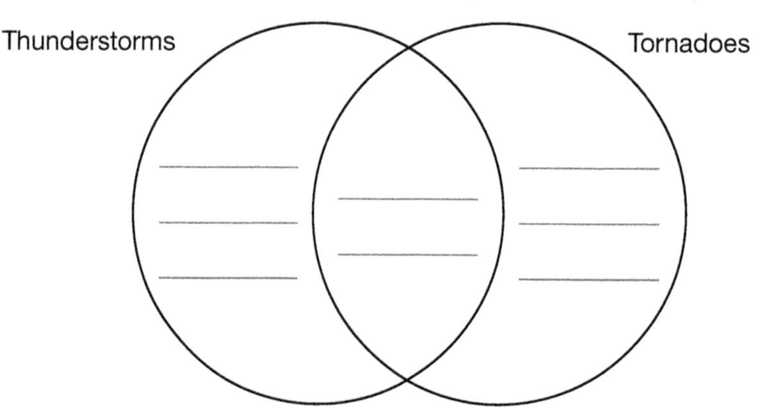

C Add at least one more fact from the reading to each circle of the Venn diagram. Then add another fact to the overlapping area.

2 Using a dictionary V

> Sometimes you will need a dictionary to help you figure out the meaning of a word in a text. Many words have more than one definition. It is important to choose the definition that is appropriate for the context in which you see the word.

Read the sentences from "Storms." Each word in **bold** has more than one meaning. Use the context and a dictionary to choose the correct meaning and part of speech. Then write the meaning and part of speech in the blank. Follow the example.

Example:
The hot air expands. Then it quickly **contracts** as it cools down.

Dictionary definition:

> contract /'kɑn·trækt/ (n) **1.** a written legal agreement between two people or companies that says what each will do **2.** an agreement to kill someone for money /kən'trækt/ (v) **1.** to get an illness. **2.** to become smaller

Correct definition: *(v.) to become smaller*

1. These **short** thunderstorms bring cool rain on a hot, humid day.
   Correct definition: ______________________
2. Approximately 90 percent of thunderstorms are small and **last** no longer than 30 minutes.
   Correct definition: ______________________
3. Every year, lightning causes **over** 7,000 forest fires, hundreds of injuries, and more than 90 deaths in the United States.
   Correct definition: ______________________
4. They are usually **over** in 15 minutes or less, but they move quickly and cause a great deal of damage in just a few minutes.
   Correct definition: ______________________
5. On average, more than 100 tornadoes **strike** Texas each year.
   Correct definition: ______________________

3 Using *this / that / these / those* to connect ideas 

Writers often use *this*, *that*, *these*, or *those*, followed by a noun or noun phrase, to help connect the ideas in sentences.

noun phrase

Hail sometimes falls during a severe thunderstorm. **These** balls of ice can hurt people and damage crops.

The phrase "These balls of ice" refers to *hail* in the preceding sentence and helps the reader understand that *hail* means "balls of ice."

Reread paragraphs 1, 2, 4, and 5 of the text. Find examples of phrases with *this*, *that*, *these*, or *those*. Write each phrase and the idea it refers to.

1. Paragraph 1
   Phrase: *This movement of air. . .*
   Refers to: *The hot air expands. Then quickly contracts . . .*
2. Paragraph 2
   Phrase: __________
   Refers to: __________
3. Paragraph 4
   Phrase: __________
   Refers to: __________
4. Paragraph 5
   Phrase: __________
   Refers to: __________

4 Examining statistics (A) (R)

College courses test understanding of statistical information. You will often find statistics that are general, that is, not exact, in texts. Instead, the words *almost*, *approximately*, *over*, and *more than* may be stated before a statistic to make it less exact.

Hail causes *almost* $1 billion of damage each year.

The word *almost* makes the statistic ($1 billion) less exact. The sentence means that hail causes less than $1 billion of damage each year, but very close to $1 billion.

A Find statistics in the reading "Storms." Underline them. Then circle the word or phrase that makes each statistic less exact.

B Now answer these questions.

1. How many thunderstorms might be happening at this moment around the world?

a. 1,000
b. 1,987
c. 2,238

2. Last week there was a tornado in Texas. How many minutes did it probably last?

a. 10
b. 25
c. 50

3. How many tornadoes will probably occur in the United States next year?

a. 100
b. 1,000
c. 1,100

4. Lightning strikes happen ________ million times every year.

a. 38
b. 41
c. 70

5. Firefighters think that lightning will cause ________ forest fires next year.

a. 6,800
b. 7,200
c. 7,000

6. How many tornadoes will probably strike Canada next year?

a. 110
b. 50
c. 1,000

PREPARING TO READ

1 Thinking about the topic (R)

Work with a partner to complete the following activities.

A The text you are going to read is about hurricanes. What are hurricanes? Look the word up in a dictionary. What kind of words and ideas do you expect to find in the reading about hurricanes?

B Did you ever experience a hurricane? Where were you? What happened? Describe your experience.

2 Increasing reading speed (R)

A Read the text "Hurricanes." Use the strategies for increasing reading speed on page 68. For this task, do not read the boxed text on page 142.

1. Before you begin, fill in your starting time.
2. Fill in the time you finished.

Starting time: ________
Finishing time: ________

B Calculate your reading speed:

Number of words in the text (493) ÷ Number of minutes it took you to read the text = your reading speed

Reading speed: ________

Your goal should be about 80–100 words per minute.

C Check your reading comprehension. Circle the correct answers. Do not look at the text.

1. Hurricanes form over (*warm waters / cold waters / land*).
2. Hurricanes have (*one / two / three*) main parts.
3. Hurricanes can cause a lot of (*noise / damage / warm water*).
4. Hurricanes can be (*deadly / costly / deadly and costly*).
5. Many scientists think there will be (*more / fewer / no*) hurricanes in the future.

Reading 3

HURRICANES

People in China call them typhoons. People in India refer to them as tropical cyclones. People in the United States call them hurricanes. Different parts of the world use different names, but they are all referring to the same thing: a very powerful, spinning storm that causes strong winds, heavy rain, and giant waves.

The formation of a hurricane

A hurricane begins near the equator, over warm tropical waters. It might start in the southern part of the North Atlantic Ocean, the Caribbean Sea, the Gulf of Mexico, or the East Pacific Ocean. The storm starts as an area of thunderstorms. It continues to grow, and the winds blow furiously in a circular path. When the winds reach 118 kilometers per hour, the storm is called a hurricane. The source of a hurricane's energy is the warm ocean water it travels over. As it moves over colder waters or over land, it weakens.

The parts of a hurricane

Hurricanes have three main parts: the eye, the eyewall, and the spiral, or curving, rain bands. The eye of a hurricane is a calm area with only light winds. The eye is in the middle of the spinning storm. The eyewall is a wall of clouds around the eye. The eyewall contains the strongest winds and the most rain. Rain bands are long strips of thunderclouds that also follow a circular path around the eye of the hurricane, but they are farther away than the eyewall. Like the eyewall, rain bands bring rain and strong winds.

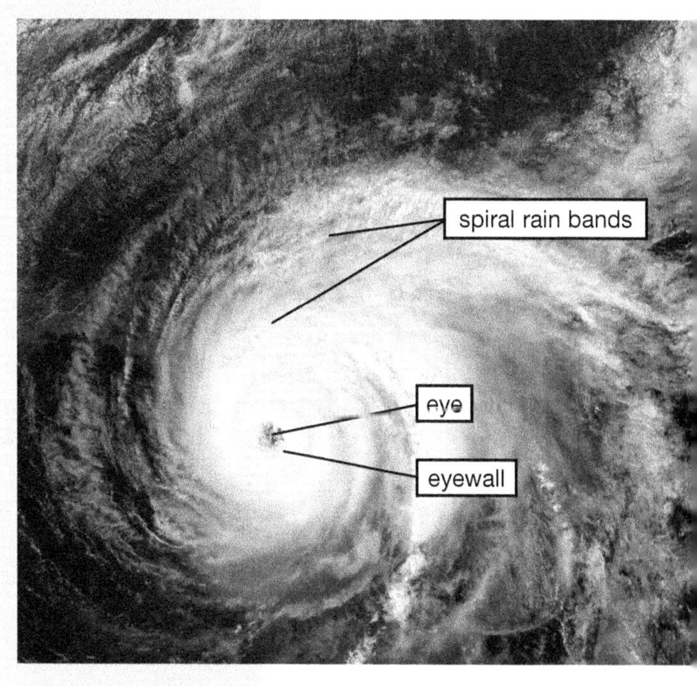

Hurricane damage

The hurricane's strong winds move it across the ocean. When a hurricane reaches land, the powerful winds, heavy rain, and **storm surge** can cause death and destruction. Hurricane winds can blow more than 320 kilometers per hour. They can knock down trees, destroy buildings, and blow heavy objects through the air. The strong winds can cause great harm, but water usually produces the most damage. More than 60 centimeters of rain can fall in one day, and the storm surge can be as high as 6 meters. All of this water can wash away beaches, roads, bridges, homes, crops, animals, and people.

storm surge a sudden rise in the ocean's water level as it moves closer to land during a storm

35 Hurricane damage can be overwhelming. In 2005, Hurricane Katrina caused more than $80 billion of damage. It killed more than 1,500 people and destroyed more than 250,000 homes in the southeastern part of the United States. Although Katrina was the costliest hurricane in United States history, the Galveston, Texas, 40 hurricane in 1900 was the deadliest. More than 8,000 people died in that storm. However, the deadliest hurricane of all was the 1970 Bangladesh cyclone. It killed more than 300,000 people and devastated the country.

Hurricanes and global warming

Hurricanes are the most destructive storms on Earth, and science 45 is searching for a way to stop them. In 2008 and 2009, some scientists suggested adding smoke, very cold water, and even a giant ice cube to hurricanes to stop them. Others recommended adding cold water on top of the ocean to weaken a hurricane. However, no one has found a way to stop them yet. Many scientists believe that ocean temperatures 50 will continue to rise with global warming, and there will be more and more hurricanes in the future. Throughout the world, people are working to improve hurricane forecasting and emergency planning and to better prepare us for these devastating storms. At the same time, people are looking for ways to slow down global warming and to 55 reduce its extremely destructive effects.

wind turbine a tall, thin, metal windmill that produces electricity

The Benefits of Wind

Hurricane and tornado winds can be destructive and deadly. However, wind can also be a positive force. In fact, some people are using the power of wind to produce energy. For example, Germany, Spain, and the United States use **wind turbines** to generate electricity in some areas, and these turbines are becoming more popular in other countries, too.

Wind energy has many advantages. It is a cheap way to produce electricity, it is a clean source of power, and it is unlimited. As long as the wind blows, people can use it to produce energy. However, wind energy has a few disadvantages. For example, wind turbines are useful only in areas with strong, regular winds, so they cannot be used everywhere. In addition, some people think they are ugly and dangerous to birds.

There may be problems, but wind power is a clean, low-cost, and renewable way to produce energy. It may also help us solve two of our planet's most serious problems: pollution and global warming.

AFTER YOU READ

1 Reading for main ideas (R)

A Look at the paragraph topics of "Hurricanes." Find the paragraph for each topic in the reading and write the number in the blank.

1. the three parts of a hurricane Par. ____
2. the connection between global warming and hurricanes Par. ____
3. different names for hurricanes Par. ____
4. the power of a hurricane Par. ____
5. the damage that specific hurricanes have caused Par. ____
6. how hurricanes form Par. ____

B Check (✓) the sentence that expresses the main idea of the whole text.

____ **1.** Scientists believe that global warming may cause more hurricanes in the future.
____ **2.** Hurricanes have spinning winds, a calm eye, and an eyewall.
____ **3.** Hurricanes are powerful storms that can cause a lot of damage and destruction.

C Compare answers with a partner.

2 Synonyms (V)

Synonyms are words that have similar meanings (for example, *big* and *large*). You can use synonyms to avoid repeating the same words. This will make your writing more interesting.

Go back to the reading and complete the following activities.

A Find the following words and underline them.

1. call (v) ________ (Par. 1)
2. begins (v) ________ (Par. 2)
3. strong (adj) ________ (Par. 4)
4. destruction (n) ________ (Par. 4)
5. destructive (adj) ________ (Par. 6)

B Find a synonym for each word in Step A in the same paragraph. Circle it. Note that the synonym will be the same part of speech. Write the synonym in the correct blank above.

C Write two sentences with one pair of synonyms from Step A. Use the second synonym to connect the meaning of the two sentences.

Last year, there was a devastating tornado in my city. The destructive storm caused hundreds of injuries.

3 Prepositions of location (V)

Prepositions of location show the spatial relationship between two or more things. For example, in the diagram on the right:

- the circle is over the square.
- the square is near the heart.
- the triangle is in the heart.
- the heart is around the triangle.

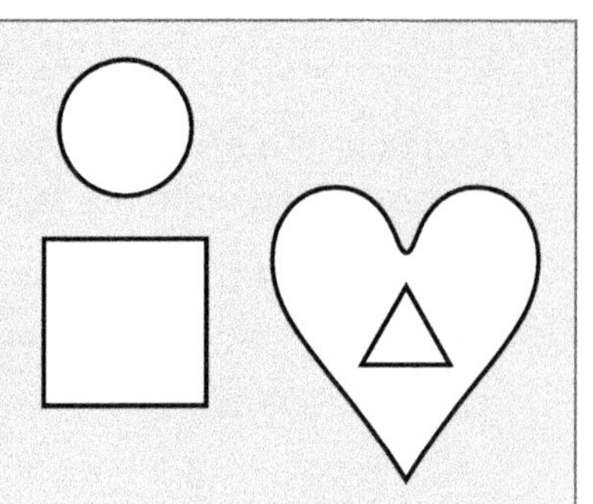

A Read the sentences. Circle the preposition of location in each sentence. Underline the two things that it connects in space. Then draw a simple diagram to show the spatial relationship between these two things.

1. Hurricanes form near the equator.

2. Some hurricanes form over the warm waters of the Gulf of Mexico.

3. The eye of a hurricane is in the middle of the spinning storm.

4. Rain bands circle around the eye of the hurricane.

B Look at the photograph below. Work with a partner and take turns describing what you see. Use prepositions that show location.

Examples:

The boat is in the street.

The cars are in deep water.

4 Thinking critically about the topic Ⓐ

> In class and on tests, you will often have to think about the ideas in a text in relation to your own knowledge and experience.

Discuss the following questions in a small group.

1. What have you noticed or heard about Earth's changing climate?
2. What are some things you can do to help slow global warming?
3. Some people, companies, and countries have made changes to help slow global warming, and others have not. Why do you think this is true?

Chapter 6 Academic Vocabulary Review

The following words appear in the readings in Chapter 6. They all come from the Academic Word List, a list of words that researchers have discovered occur frequently in many different types of academic texts. For a complete list of all the Academic Word List words in this chapter and in all the readings in this book, see the Appendix on page 206.

approximately	contracts (v)	global	period
benefits (n)	expand	injuries	positive
consists	generate	located	research (n)

Complete the sentences with words from the list.

1. Flooding and drought are ________ problems. Almost every country has to deal with issues of too much water and not enough water.
2. Houses that are ________ in Tornado Alley may get damaged every year.
3. One of the ________ of thunderstorms is the water they bring to the land.
4. Scientists often publish the results of their ________ so that other people can learn about their discoveries.
5. When you take a breath of air, your lungs ________.
6. More people should wear seat belts to prevent traffic ________.
7. There are ________ 7.042 billion people in the world.
8. The most common ________ of time for Atlantic hurricanes is June through November.
9. A balanced diet ________ of protein, whole grains, dairy, fruits, vegetables, and healthy fats.
10. Hurricanes often ________ a dangerous storm surge. This is why people should stay away from the ocean during a hurricane.

Practicing Academic Writing

In Unit 3, you learned about Earth's atmosphere and weather conditions. Based on everything you read and discussed in class, you will write a paragraph about this topic.

Climate

You will write one academic paragraph about the climate in a place you know. Describe the climate in as much detail as you can. Discuss the weather conditions that are common in the area.

PREPARING TO WRITE

1 Using a cluster diagram as a prewriting strategy

Using graphic organizers is often a good way to take notes on a reading. A graphic organizer, such as a cluster diagram, can also help you generate ideas on a topic.

Cluster diagrams "cluster" your ideas, information, and details around a topic. They can help you prepare to write because they help you develop ideas. When you are finished diagramming, they will also show you which key points have the most support and which might be good points to include in a paragraph.

Follow these steps to create a cluster diagram.

1. Write your topic in the largest circle.
2. Write key ideas about the topic in the smaller circles.
3. Write details that support the key ideas in the smallest circles.

Think fast and carefully and jot your ideas down. You can add more ideas by adding circles.

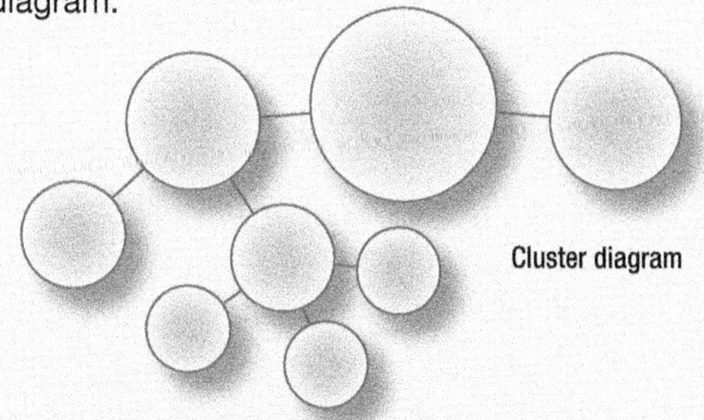

Cluster diagram

A Choose the place you are going to write about. Think about your favorite places or places that you did not like. The place can be from your past or present, such as your home now, an old home, or a place that you visited or like to visit.

B Brainstorm your topic. Explore the climate of the place you chose with a cluster diagram. Get a separate piece of paper. Write your topic in a large circle in the middle of the paper. Add the major climate details in smaller circles around the topic. Think about temperature, precipitation, seasons, or storms, for example. Next, you need to add specific support.

2 Using specific support in your writing

You learned in this unit that examples and statistics can be an effective way to include specific support. Examples and statistics make writing more interesting and "real." They help prevent writing from becoming too general and vague. When you use statistics, you don't always have to give exact numbers. It is usually easier for a reader to understand a text with *rounded* numbers. You can round numbers up or down to the nearest whole number.

Examples:

- rounding up: 46.7 → 47
- rounding down: 46.3 → 46

A Read the following information about one type of powerful storm.

Nor'easters are powerful storms. Winds that blow from the northeast create these storms. Nor'easters can be very dangerous and can cause many problems. Some nor'easters become so famous that people talk about them for many years.

B Notice that the paragraph is very general. It lacks specific examples and statistics.

C Read the details below.

Nor'easters

- occur along the East Coast of the United States.
- occur between October and April.
- can bring heavy snow, rain, winds, and giant waves with flooding.
- can cause serious damage.
- can kill people.

The Blizzard of 1978

- caused $502 million of damage.
- lasted 32 hours and 40 minutes.
- damaged 9,406 houses.
- destroyed 2,163 houses.
- killed 73 people in Massachusetts.
- dumped 66 centimeters of snow on the city of Boston, Massachusetts.
- forced 17,008 people to go to emergency shelters.

D Revise the paragraph in Step A. Include specific examples and statistics from Step C. Remember to do the following:

- use numbers, words, and phrases to make the statistics less exact (*almost*, *approximately*, *over*, *more than*)
- use *for example* and *such as* to introduce examples
- use your own words to state your information

E Go back to the cluster diagram that you created for your paragraph in Task 1, Step B. Does it include specific details, examples, and statistics? Brainstorm specific details now and write them down in your diagram. Find additional details in the library or on the Internet as needed.

NOW WRITE

A Now write the first draft of your paragraph.

B Start with a clear topic sentence. Include at least three major supporting details. Use examples and statistics to explain and support your ideas. Make sure you add specific supporting examples and statistics. Use this checklist.

Are you including:

____ a topic sentence that states the main idea of the paragraph

____ major supporting details

____ minor details that illustrate the major or key support

____ a concluding sentence that restates the main idea (Be sure to make the concluding sentence a little different from the topic sentence.)

____ correct paragraph form and structure

____ vocabulary you learned in this chapter

____ correct sentences with subjects and verbs that agree

C Give your paragraph a title.

AFTER YOU WRITE

A Exchange paragraphs with a partner and read each other's work. Then discuss the following questions about both paragraphs:

- What is the most interesting information in your partner's paragraph?
- Does your partner's paragraph have correct form and structure? How do you know? Explain.
- Do the topic sentences make a claim about the topic?
- Are supporting details included, such as examples or facts?
- Are all of the ideas presented in a logical order? What are the transition words?
- Is all of the information on topic?
- Are there spelling or grammar mistakes?

B Think about any changes to your paragraph that would improve it. Then write a second draft of the paragraph.

Unit 4

Life on Earth

In this unit, you will look at living things in the natural world. In Chapter 7, you will discuss the features that all living things share. You will also examine plant and animal life. In Chapter 8, you will focus on parts of the human body: the brain, the skeletal and muscular systems, and the heart and the circulatory system.

Contents

In Unit 4, you will read and write about the following topics.

Chapter 7 Plants and Animals	Chapter 8 Humans
Reading 1 Living Things **Reading 2** Plant Life **Reading 3** Animal Life	**Reading 1** The Brain **Reading 2** The Skeletal and Muscular Systems **Reading 3** The Heart and the Circulatory System

Skills

In Unit 4, you will practice the following skills.

 Reading Skills	 Writing Skills
Thinking about the topic Building background knowledge about the topic Previewing key parts of a text Applying what you have read Increasing reading speed Asking and answering questions about a text Scanning for details Sequencing	Writing about similarities Writing about differences Writing about similarities and differences Writing a description Writing about the body
 **Vocabulary Skills**	 **Academic Success Skills**
Word families Defining key words Cues for finding word meaning *That* clauses Compound words Using adjectives Gerunds Using a dictionary Words that can be used as nouns or verbs Prepositions of direction Playing with words	Answering true / false questions Asking for clarification Conducting a survey Making an outline Thinking critically about the topic Highlighting and taking notes Conducting an experiment Answering multiple-choice questions Highlighting and making an outline

Learning Outcomes

Write an academic paragraph about the human body

Previewing the Unit

Previewing means looking at one thing before another. It is a good idea to preview your reading assignments. Read the contents page of every new unit. Think about the topics of the chapters. You will get a general idea of how the unit is organized and what it is going to be about.

Read the contents page for Unit 4 on page 152 and do the following activities.

Chapter 7: Plants and Animals

A Go outside or look out the window. What do you see? Make a list of the living things. Then make a list of things that are not alive. Compare lists with a partner. What do you think all living things have in common?

B Read these features of plants and animals. Label each feature *P* (plants) or *A* (animals). Then add one feature of plants and one feature of animals.

____ **1.** have leaves
____ **2.** have a brain
____ **3.** raise their young
____ **4.** can make their own food
____ **5.** can communicate with each other
____ **6.** produce oxygen

Chapter 8: Humans

A This chapter focuses on various parts of the human body. How much do you know about the body? Read the statements. Write *T* (true) or *F* (false).

____ **1.** The heart is a muscle.
____ **2.** The brain is a bone.
____ **3.** The brain is bigger than the heart.
____ **4.** Bones are heavy.
____ **5.** There are 105 muscles in the body.
____ **6.** The rib bones protect the heart.

B Look at the pictures below. Which one shows the skeletal (bone) system? Which one shows the muscular system? Label each picture with the correct name.

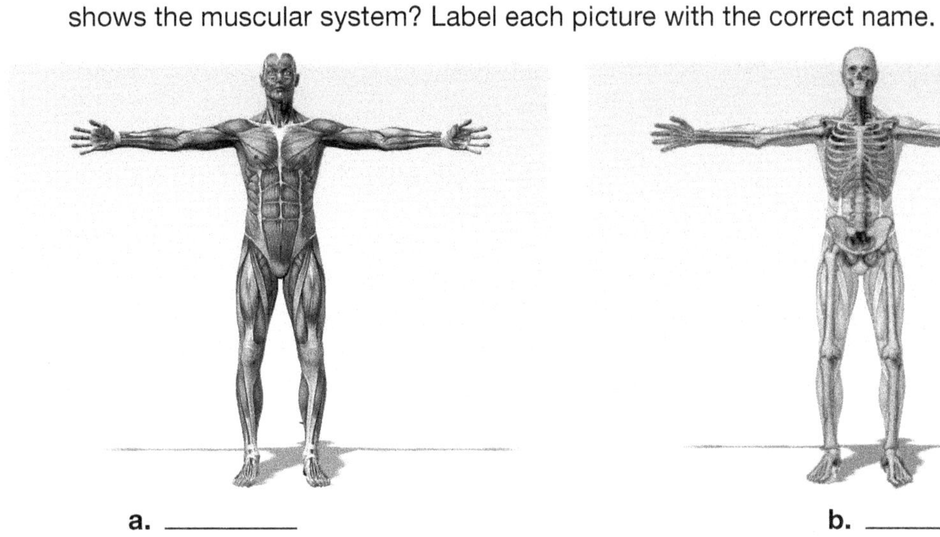

a. __________

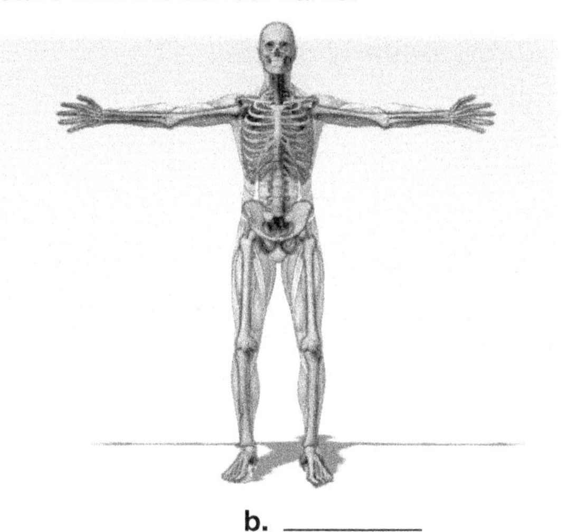

b. __________

Chapter 7
Plants and Animals

PREPARING TO READ

1 Thinking about the topic Ⓡ

A Read the sentences below and check (✓) the ones that are true.

_____ **1.** All living things need water.
_____ **2.** All living things move from place to place.
_____ **3.** All living things need a place to live.
_____ **4.** All living things need air to breathe.
_____ **5.** All living things need food.
_____ **6.** All living things communicate with each other.
_____ **7.** All living things grow.
_____ **8.** All living things die someday.

B Compare your answers with a partner.

C With your partner, add two more true sentences to the list. Share your ideas with the class.

2 Building background knowledge about the topic Ⓡ

A Read the following information about cells.

> All living things are made up of cells. A cell is the smallest unit of life. All cells have an outer covering called a **cell membrane.** Inside every cell is a jellylike material called **cytoplasm.** Most cells also have a **nucleus** in the middle of the cytoplasm.

B Look at the diagram of a cell below and label the three main parts: *cell membrane*, *cytoplasm*, and *nucleus*.

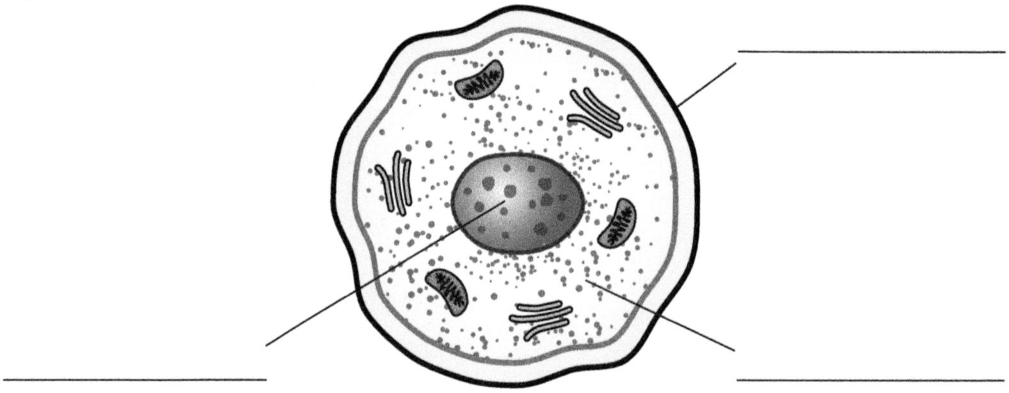

____

______________________ ______________________

Reading 1

LIVING THINGS

What is the difference between a rock on the ground and a plant that grows next to it? What is the difference between the ocean and a fish that lives there? What is the difference between a cloud in the sky and a bird that flies through it? What do plants, fish, and birds have in common? They are all **organisms**, or living things. Rocks, the ocean, and clouds are inorganic, or nonliving, things.

organism a living thing

Life on Earth is extremely diverse, or full of many different types of things. When you look around, you notice that organisms are many different shapes and sizes. Some, like ants, are very small. Others, like whales, are quite large. Organisms also live in a variety of places. Some live in the air, others live in or on the earth, and many live in the ocean. Although organisms are different from each other in these ways, they are similar in other ways. For example, they all need water, food for energy, and a place to live. In addition, all organisms grow, develop, and eventually die.

cell the smallest unit of life

microscope a scientific instrument that makes very small objects look bigger

Another important similarity is that all organisms are composed of **cells**. A cell is the smallest unit of life. Most cells are so small that you can only see them with a **microscope**. All cells have an outer covering, called a cell membrane. This membrane keeps the cell material inside 20 the cell. It also controls the movement of things into and out of the cell. For example, the membrane allows water to move into and out of a cell, but it does not allow dangerous materials to enter. Inside every cell is a jellylike material called cytoplasm. Most cells also have a nucleus in the middle of the cytoplasm. The nucleus controls all 25 activity in the cell.

Most organisms on Earth are made up of just one cell. For example, both bacteria and algae are single-celled organisms. Other, larger organisms, such as some plants and all humans, are multicellular organisms. They are made up of many cells. In fact, the human body is 30 made up of billions of cells.

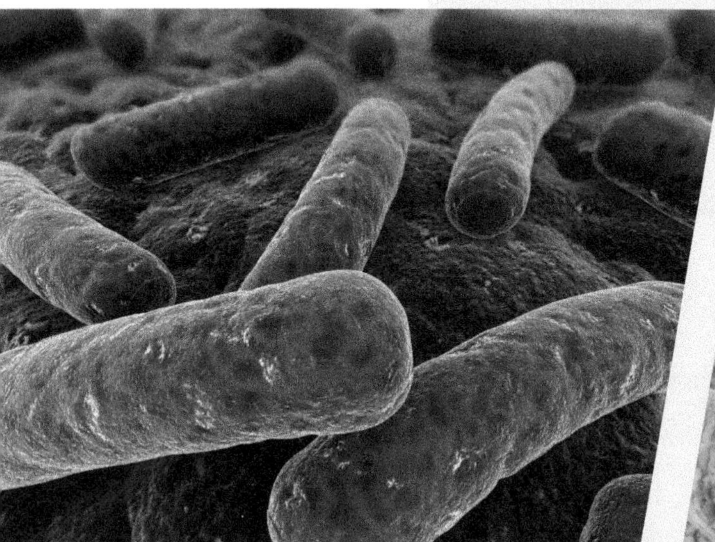

Bacteria cells

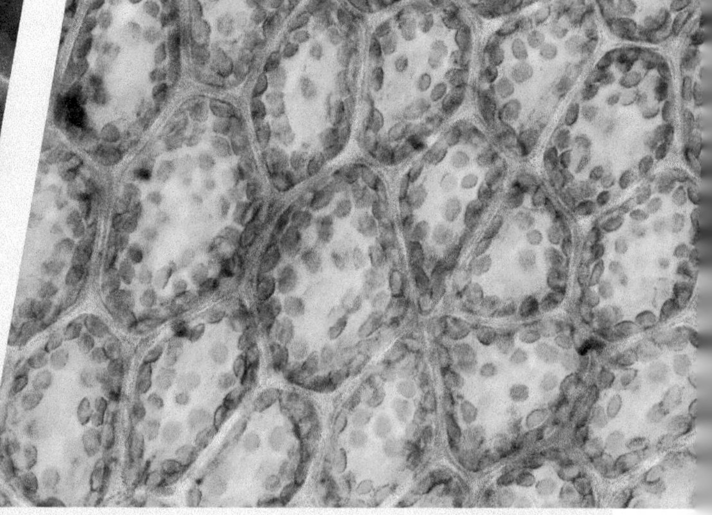

Leaf cells

The cells in multicellular organisms work together. A group of cells that work together to do one job is called tissue. For example, one group of thousands of cells forms muscle tissue in the body. When two 35 or more different types of tissue work together, they form an organ, such as a person's heart, an elephant's ear, or a plant's leaf. When a number of organs work together, they form an organ system, such as the circulatory system (the system that transports food and gases through the body with the help of the heart and lungs).

Biologists (scientists who study living things) continue to learn 40 more and more about the organisms on our planet. Some of their research findings may help us protect the rich diversity of life on Earth in the future.

AFTER YOU READ

1 Answering true/false questions A R

A Review the strategies for answering true/false questions on page 36. Read the statements below and find the answers in the reading. Answer *T* (true) or *F* (false). Then write the number of the paragraph where you found the information.

____ **1.** Some organisms need water. Par. ____
____ **2.** Organisms grow and die. Par. ____
____ **3.** Humans, animals, and plants have organs. Par. ____
____ **4.** Bacteria and humans are multicellular organisms. Par. ____
____ **5.** All cells have cell membranes, cytoplasm, and a nucleus. Par. ____
____ **6.** Rocks, fish, and birds are all organisms. Par. ____
____ **7.** You often need a microscope to see a cell. Par. ____
____ **8.** Some organisms are made up of many cells. Par. ____

B Work with a partner. Correct the sentences you marked false in Step A.

2 Word families V

When you learn a new word, try to learn some other words in its word family as well. This will develop your vocabulary. For example, when you learn the verb *destroy*, you can also learn the related words *destruction* (n) and *destructive* (adj).

A Look at each word below. Find a related word in the reading. Write the parts of speech next to the words. Use the abbreviations *n*, *v*, and *adj*.

1. life _(n)_ _living (adj)_
2. difference ____ ________
3. similarity ____ ________
4. cell ____ ________
5. movement ____ ________
6. diversity ____ ________

B Complete each sentence below with an appropriate word from Step A.

1. Fish and birds live in ________ places.
2. Bacteria are made up of just one ________.
3. Many materials ________ into and out of a cell.
4. Plant cells are ________ to animal cells. They have many things in common.
5. There is a lot of ________ on Earth. There are many different kinds of organisms.

3 Asking for clarification Ⓐ

As a student, you will have many occasions when you do not understand a word, an idea, or what your instructor says. When you are not sure what something means, you can ask questions to find out. This is asking for clarification.

To **clarify** means to restate something or to say it again so people understand. To ask for clarification means to ask someone to restate or to say something again so that you or other people can better understand.

Here are some structures you can use to ask for clarification:

- I'm sorry. I don't understand. Can you say that again?
- I'm not sure what the author means when she says that . . .
- Could you explain what the word __________ means?
- Could you give me an example of __________?
- I don't understand what the text means in paragraph _____. It says . . .

A Read the following clarification questions and statements about the reading.

1. I'm not sure what the author means in paragraph 3 when she says that a cell is the smallest unit of life.
2. Could you explain what the word *circulatory* in paragraph 5 means?
3. Could you give me an example of another organ system in addition to the circulatory system?
4. I don't understand what the text means in paragraph 2 where it says that all organisms develop.

B With a partner, answer the clarification questions in Step A.

C Find other words and ideas in the reading that you would like to clarify. With your partner, take turns asking and answering clarification questions.

4 Writing about similarities W R

In academic texts, writers often use certain expressions to compare two or more people, places, things, or ideas. These expressions show how things are similar (alike in some way). Look at the examples below.

Here are some ways to compare two things:

- Humpback whales and bottlenose dolphins **are similar (to each other)** in many ways.
- They **have** many features **in common**.
- **Both** humpback whales and bottlenose dolphins live in the ocean.
- **One similarity is that** they **are both** mammals.
- **Another similarity is** that they **both** eat fish.

Here are some ways to compare more than two things:

- Whales, walruses, and seals **are similar (to each other)** in many ways.
- They **have** many features **in common**.
- **One similarity is that** they **are all** mammals.
- **All** mammals **have** body temperatures that stay about the same.

A Go back to the reading. Underline the expressions of similarity.

B Complete these expressions of similarity. Use the structures in the strategies box as a guide.

1. Roses, tulips, and orchids are __________ in many ways. One __________ is that they are __________ popular flowers.
2. Dogs and wolves are __________ natural hunters.
3. __________ tissues and organs are made of cells.
4. __________ living things need food to eat and a place to live.

C Write three sentences of your own. In each sentence, compare two animals and identify one similar feature between them. Compare your sentences with a partner.

1. ____

____

2. ____

____

3. ____

____

D Read the following paragraph and discuss the questions below.

Lions and tigers are similar in three ways. One similarity is the food they eat. Lions and tigers are both meat eaters that hunt other large and medium-size animals. Another similarity is that both animals are part of the cat family. Lions are also similar to tigers in size. Both animals are about the same weight and height.

1. What two things is the writer comparing?
2. How many points of comparison does the writer discuss? What are they?
3. Underline all the expressions of similarity in the paragraph.

E Now write a paragraph about the similarities of dogs and cats. Remember to start with a topic sentence. Include points that the animals have in common and use some expressions of similarity. End with a concluding sentence.

PREPARING TO READ

1 Conducting a survey A

The text you are going to read is about plants. Plants are an important part of life on our planet. They give us oxygen and food, add beauty to our world, and provide the materials for many products.

A Survey five people for information about plants. First copy this chart. Then ask each person the questions in the chart. Record their names and answers.

Name	Why are plants important?	What are three products we get from plants?
1.		
2.		
3.		
4.		
5.		

B Share your survey results with the class.

2 Previewing key parts of a text R

A Preview the title, headings, and photographs on pages 162–163. This will give you a general idea of what the text is about. Then read the first sentence of each paragraph.

B Now read the statements below and answer true (*T*) or false (*F*).

____ **1.** There are thousands of different types of plants on Earth.

____ **2.** There are three general categories of plants.

____ **3.** Earth has more seed plants than seedless plants.

____ **4.** We could survive on Earth without plants.

____ **5.** Plant life on Earth is decreasing.

Reading 2

PLANT LIFE

The diversity of plants

There are approximately 300,000 types of plants on Earth. Plants grow almost everywhere. For example, moss and lichen grow in cold polar climates. Palm trees and orchids grow in hot, wet, tropical climates. Cacti and ocotillos grow in warm, dry climates. Plants are similar to other organisms in several ways, but they also have their own special features.

Plant size and structure

Just like other organisms, plants come in many different sizes. There are very small plants and very large ones. For example, you can only see tiny ferns with a microscope; in contrast, you cannot see the top of some of the giant redwood trees in northern California that grow to more than 100 meters. Plants also have some structural features in common with other organisms. Like other living things, plants are made up of cells that have a cell membrane, cytoplasm, and a nucleus. One difference is that plant cells also have a thick, rigid cell wall that surrounds the membrane. These cell walls give the plant structure and support, and allow it to grow straight and tall. Cell walls also help protect the cells.

Seedless plants and seed plants

Scientists divide plants into two categories: seedless plants and seed plants. Seedless plants grow from **spores**, not seeds. Spores are tiny cells that grow into new plants. Seedless plants do not have flowers, and most grow in places that are damp (a little wet). Mosses and ferns are examples of seedless plants.

spore a cell produced by seedless plants and some other organisms that is able to grow into a new plant or organism

Seed plants are much more common than seedless plants. All seed plants have roots, a stem, and leaves. Unlike seedless plants, many also have flowers. The flowers produce the seeds, and these seeds grow into new plants. The roots take in water from the soil. The water then travels through the stem to the leaves. The leaves take in sunlight and carbon dioxide from the air. In this way, seed plants get the materials they need to make their own food. Roses and sunflowers are examples of seed plants.

Ferns

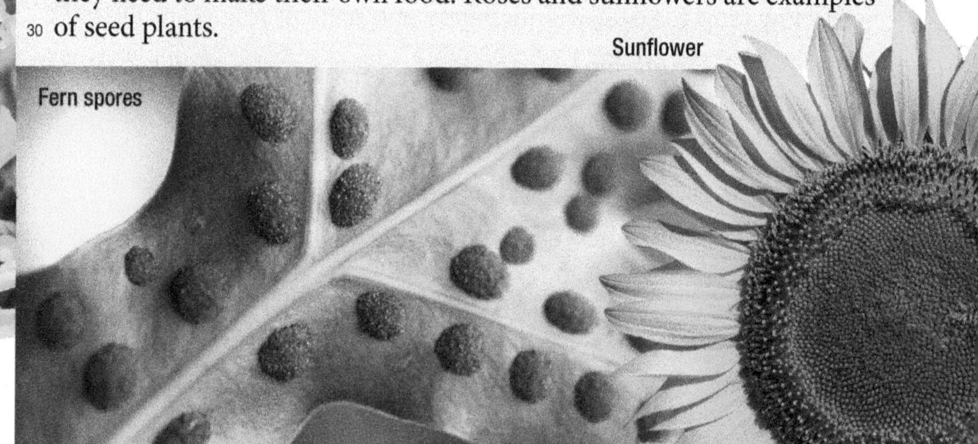

Fern spores

Sunflower

Plants as providers

Plants give us many of the things that we need to live. They provide food, clothing, paper, and wood. Some plants even function as natural medicines. However, the most important thing plants provide is oxygen, which all organisms need for survival.

35 Plants add oxygen to the air through their food-making process, called **photosynthesis**. Photosynthesis takes place in the leaves of plants. It works this way: Plants take in sunlight, carbon dioxide, and water to make their food, called glucose. Glucose is a kind of sugar. This process also creates oxygen. However, plants do not need as 40 much oxygen as they produce. Therefore, they release the oxygen into the air. When plants take carbon dioxide out of the air and put oxygen into it, they create air we can breathe. In fact, one acre of trees in a forest releases enough oxygen in a year to keep 18 people alive.

photosynthesis the process by which a plant uses energy from the sun to make its own food

Plant loss

45 Plants are essential for all life on Earth, but many plants die each year. Natural disasters, such as fires, and human activities have destroyed approximately 50 percent of the world's forests. Also, today, there is a growing demand for land. People want more building materials 50 and more space to raise animals, grow crops, and build houses. As a result, many trees are cut down. There are several negative effects of deforestation, or destruction of trees on wide areas of land. First, large numbers of plants and animals lose their habitats. In addition, scientists believe that deforestation can cause drought (a long period 55 of time without rain). Moreover, fewer trees means less oxygen in the air for us to breathe. There is also more carbon dioxide, and this contributes to global warming. For all of these reasons, our green Earth is turning browner every day.

Dangers in the Garden

Plants are very useful to humans. We eat them, build homes with them, and enjoy their beauty. However, plants can also cause harm.

Since plants cannot run away from a predator, they need ways to defend themselves. Some plants, such as rose bushes and cacti, have thorns. The thorns on these types of plants can poke, scratch, or cut people. Other plants, such as poison ivy and stinging nettles, irritate the skin. These plants can cause a red, itchy rash if someone touches them. Toxic or poisonous plants can cause dangerous, and even fatal, reactions. Many plants have poisonous leaves, stems, or flowers that negatively affect the human body and can cause death, sometimes very quickly. Oleander, deadly nightshade, and castor bean plants are examples of these types of plants.

As helpful and as beautiful as plants are, they are not always harmless.

AFTER YOU READ

1 Making an outline Ⓐ Ⓡ

An outline is a very good way to organize notes. Outlines can make ideas clear, easy to read, and easy to review. The numbers and letters of an outline show the relationships between the different parts of the text.

Complete the outline below with information from the reading. Notice that Roman numerals (*I*, *II*, *III*, etc.) introduce the main ideas, and capital letters (*A*, *B*, *C*, etc.) represent the main details.

I. Diversity of plants
   A. ≈ 300,000 types of plants on Earth
   B. grow in lots of different ______________
II. Plant size and structure
   A. different sizes
   B. plant structure: made up of ______________, which have cel ______________, ______________, a ______________, and cell ______________
III. Seedless plants
   A. grow from ______________, not seeds
   B. do not have flowers and most grow in ______________ places
IV. ______________
   A. more common than seedless plants
   B. they have ______________, ______________, and ______________, and they can have ______________
V. Products from plants
   A. provide us with food, clothing, paper, wood, and medicine
   B. most importantly, they provide ______________
VI. ______________
   A. takes place in plant leaves
   B. the process: plants take in ______________, ______________, and ______________, and they make ______________
   C. process puts ______________ into the air and takes out ______________
VII. Plant loss
   A. causes: ______________________________
   ____
   B. effects: ______________________________
   ____

2 Defining key words Ⓥ

A Match the words and definitions.

_____ **1.** stem	**a.** the parts of a plant that get water from the soil
_____ **2.** oxygen	**b.** the part of a plant that is related to making seeds
_____ **3.** roots	**c.** the gas that a plant releases into the air
_____ **4.** carbon dioxide	**d.** the part of a plant that carries water from the roots to the leaves
_____ **5.** leaves	**e.** the gas that a plant takes from the air
_____ **6.** flower	**f.** the parts of a plant that make food from water, the sun's rays, and carbon dioxide

B Label the plant diagram with the key words from Step A.

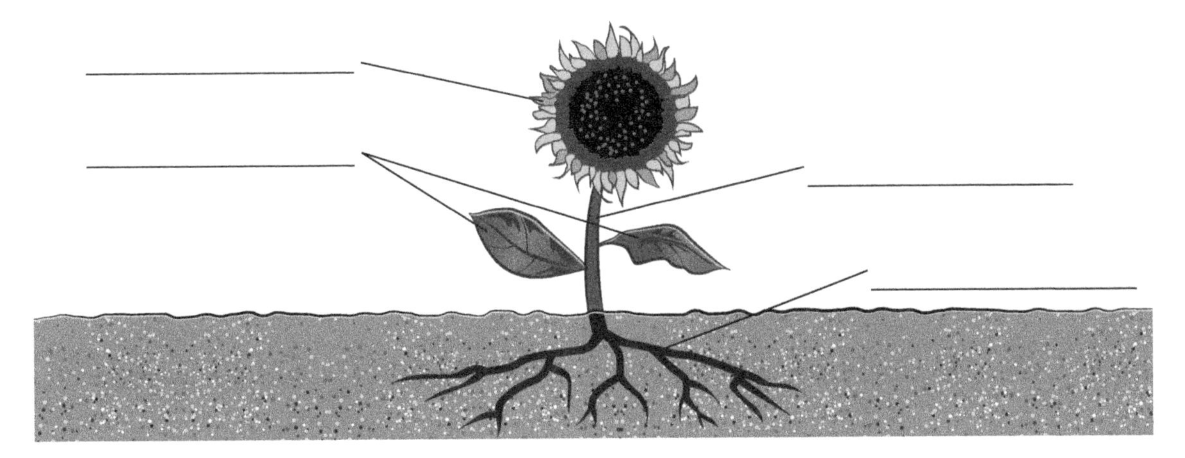

C Add the following words to the diagram below: *sunlight*, *water*, *carbon dioxide*, and *oxygen*. What process does the diagram show? Write the name of the process under the diagram.

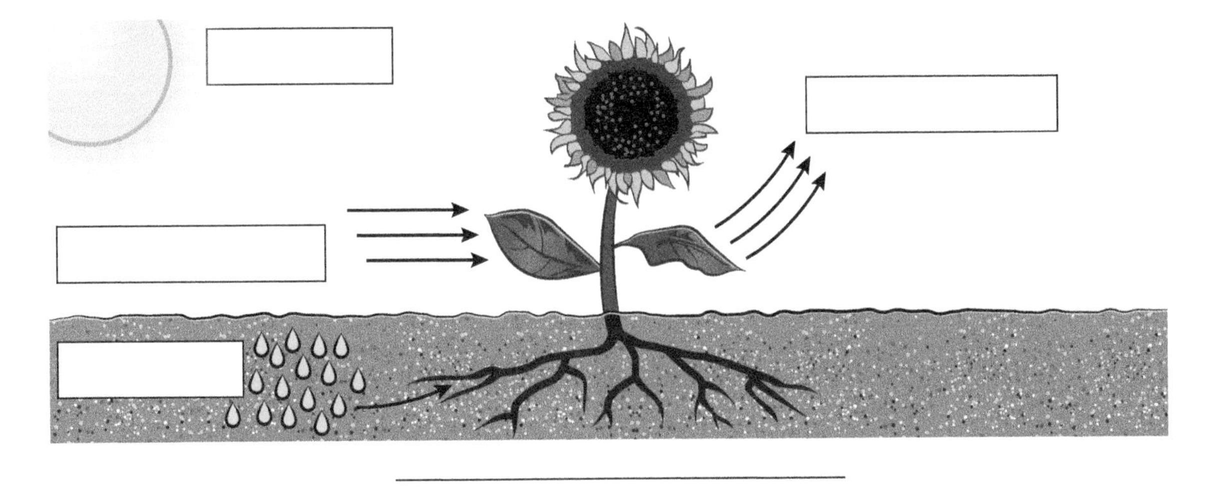

____

D Look back at the reading to check your work.

3 Cues for finding word meaning Ⓥ

Remember that the definition of a new word is often in the text. Learn to recognize different kinds of clues, and you will find the definitions. You already know some of these clues:

_________ (definition)

_________, that is, _________

_________, or _________

Other clues may be the verb *be* and the word *called*:

- A root **is** a plant part that takes in water from the soil.
- Healing plants **are** plants that people use to treat medical conditions.
- Every cell has a jellylike material inside, **called** cytoplasm.

Look at the words in the chart below. Find these words and their definitions in the reading. Then fill in the chart. Write the definition of each word and the clue that helped you find it.

Word	Definition	Clue
1. spores (Par. 3)		
2. photosynthesis (Par. 6)		
3. glucose (Par. 6)		
4. deforestation (Par. 7)		
5. drought (Par. 7)		

4 Writing about differences Ⓦ

Studies of subjects often need to contrast people, places, things, or ideas. In academic texts, you will often see structures that indicate contrast or difference. Look at these examples:

- Sunflowers grow in dry places. **In contrast**, ferns grow in places that are damp.
- **One difference between** sunflowers and ferns is that sunflowers grow from seeds, **but** ferns grow from spores.
- **Unlike** ferns, sunflowers have flowers.
- Sunflowers and ferns **are (very) different** types of plants.

A Find the expressions of difference in the reading. Underline them. Compare your answers with a partner.

B Complete the expressions of difference in the sentences below. Use the structures in the box above as a guide.

1. Mosses and palm trees are very __________ types of plants.
2. Mosses grow in cold climates. _____ __________, palm trees grow in tropical climates.
3. One __________ between moss and palm trees is their height. Palm trees are tall, but moss is short.
4. Palm trees absorb water through roots, __________ moss does not.
5. __________ moss, palm trees grow from seeds.

C Choose two different plants. Write three or four sentences that contrast the plants. Use the structures in the box. Compare your sentences with a partner's.

1. ____

____

2. ____

____

3. ____

____

4. ____

____

D Read the following paragraph. Then discuss the questions below.

There are several differences between the cacao tree and the plant called deadly nightshade. One difference is the places where they grow. Cacao trees grow in Central and South America. In contrast, nightshade grows in parts of Europe, Africa, Asia, and North America. The cacao tree and nightshade also look very different. The cacao tree grows about eight meters high, but nightshade grows only to about one meter. The most important difference between the cacao tree and nightshade is in the fruit they produce. Cacao trees produce huge berries called cacao pods. Inside the pods are the seeds that are used to make chocolate. Unlike the cacao fruit, nightshade berries are poisonous and can be fatal when eaten. As you can see, nightshade and the cacao tree are two very different plants.

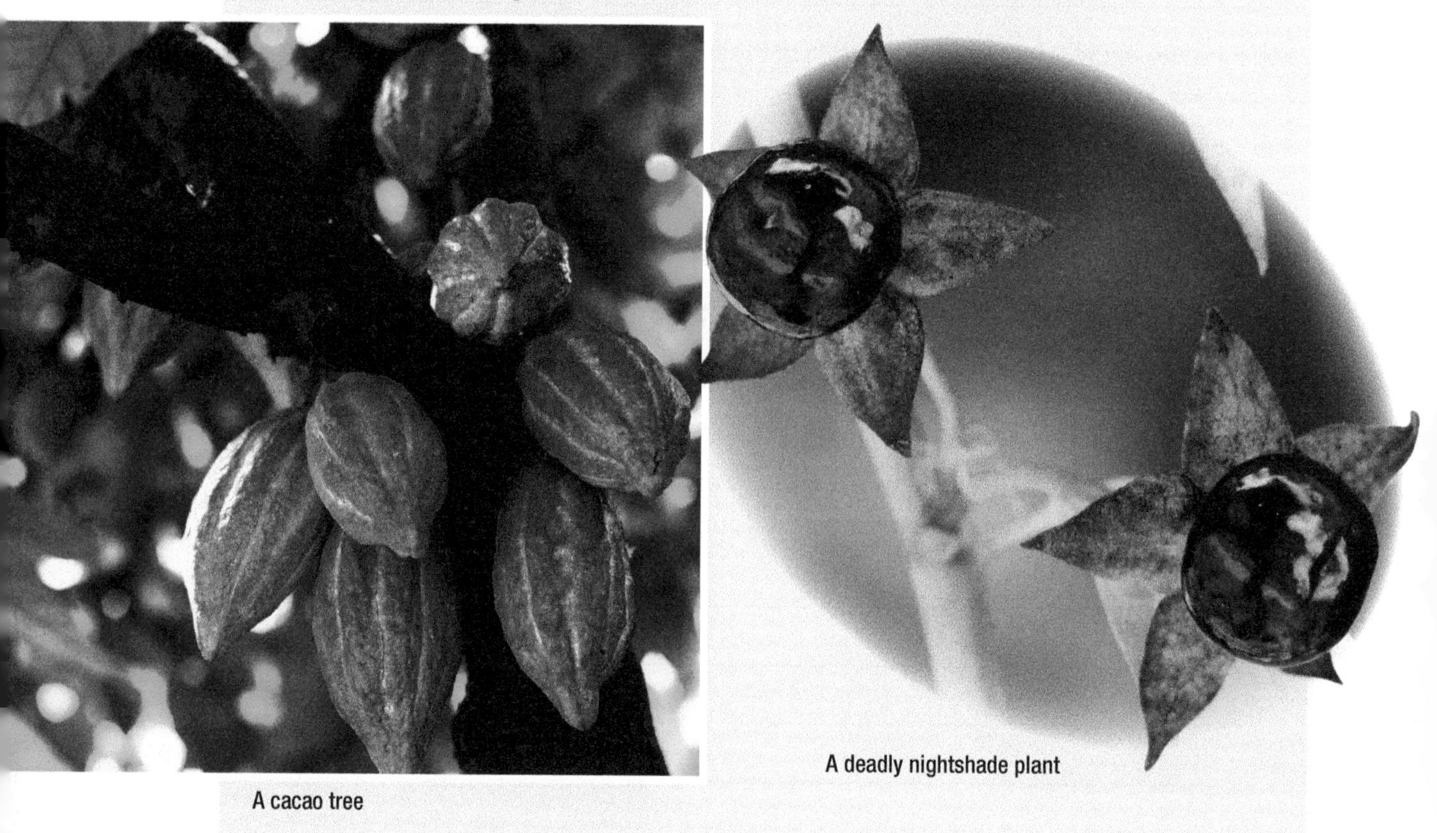

A deadly nightshade plant

A cacao tree

1. What two things is the writer contrasting?
2. How many points of contrast does the writer discuss? What are they?
3. Underline all the expressions of difference.

E Write about the differences between two plants you know. You may use the two plants you chose in Step C or other plants. Start with a topic sentence. Describe two or three differences between the plants. Use some expressions of difference that you learned in this chapter. End with a concluding sentence. Now you have a paragraph. Compare it with a partner's.

PREPARING TO READ

Thinking about the topic

The text you are going to read is about animals. There are many types of animals on our planet. Scientists group them in a variety of ways.

A Read the names of the animals in the box. Think about their similarities and differences.

bird	elephant	lion	spider
cow	fish	monkey	turtle
crab	kangaroo	mosquito	whale
dog	ladybug	mouse	worm

1. Work with a partner or small group. Divide the animals into groups. Use a dictionary for help if necessary. Put your list on a separate piece of paper.
2. Look again at the list of animals in the box. Group the animals again, but use different categories this time.
3. Discuss your choices for groups with the class. What basis did you use for the groupings?

B Look at the photographs. Notice that each one shows a relationship between two types of organisms. Then discuss the questions below in a small group.

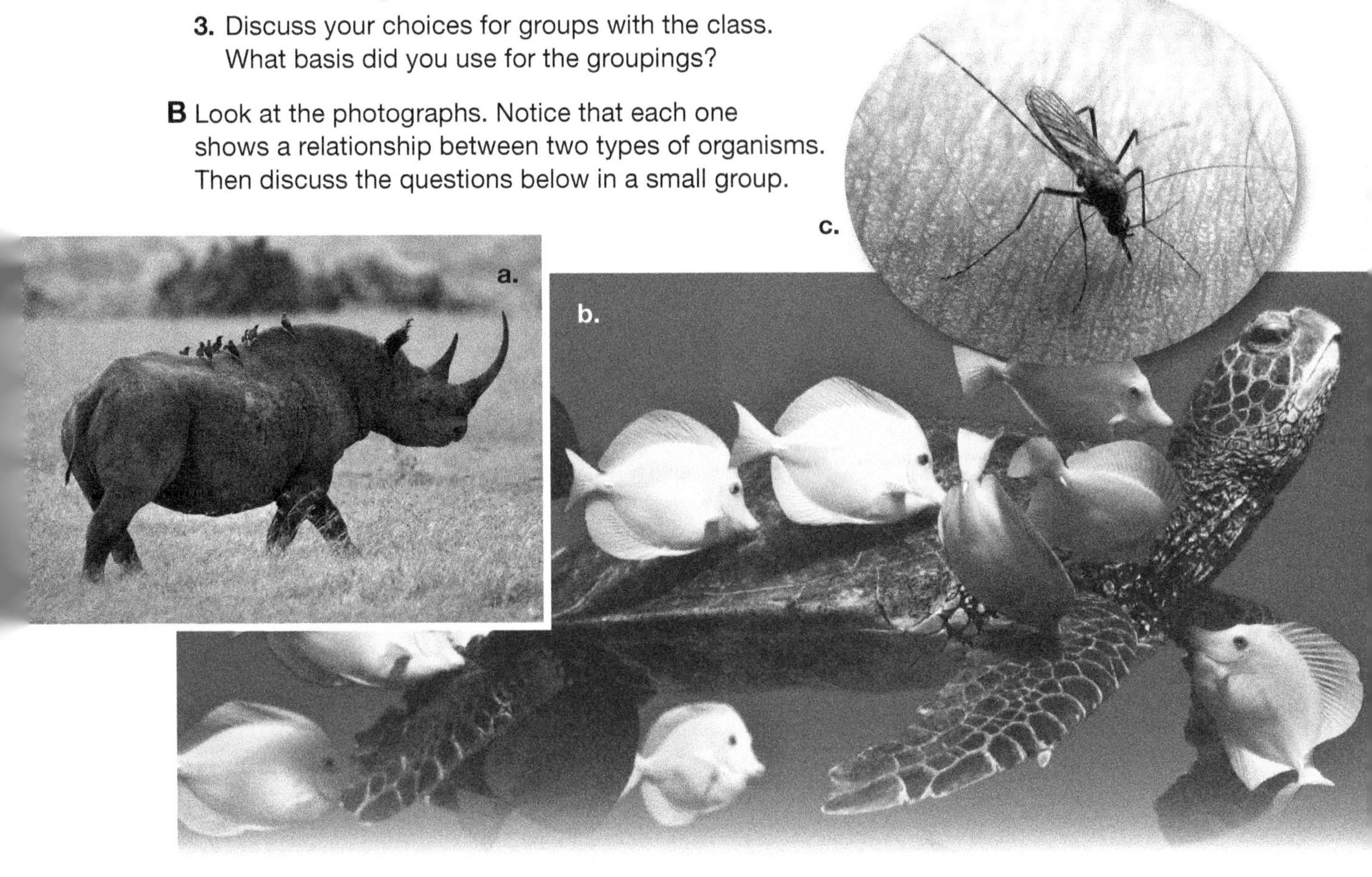

1. In photograph a, what kind of relationship do the rhinoceros and the birds have?
2. In photograph b, what kind of relationship do the turtle and the fish have?
3. In photograph c, what kind of relationship do the mosquito and the human have?

Reading 3

ANIMAL LIFE

Animal life on Earth is very diverse. There are even more kinds of animals than there are plants. Scientists have identified and named more than 1.8 million **species** of animals. They believe there are still millions more to identify in the future.

species a group of animals or plants that have the same main features and can produce animals or plants like themselves

Vertebrates and invertebrates

5 There are two main groups of animals: vertebrates and invertebrates. Vertebrates are animals that have a backbone. A backbone is a line of bones that goes down the middle of the animal's back. The backbone supports the animal. It also protects the spinal cord. The spinal cord is an important group of nerves that send messages between the brain 10 and the rest of the body. Every vertebrate also has a head with a skull. The skull surrounds and protects the brain. Fish, snakes, birds, and monkeys are all vertebrates.

Invertebrates are animals that do not have backbones, such as worms and spiders. About 95 percent of all animals are invertebrates. 15 Many of them have a hard, protective covering, such as a shell. Invertebrates are found everywhere, but most, like the starfish and the crab, live in the ocean.

Snakes are vertebrates.

Starfish are invertebrates.

Symbiotic relationships

Animals connect with each other in various ways. One way is by forming relationships, called symbiotic relationships, with other 20 organisms. Symbiosis is any close relationship between living things. There are three categories of symbiotic relationships: mutualism, commensalism, and parasitism.

Mutualism is when both organisms benefit from the relationship. For example, some small birds sit on water buffaloes and eat the insects 25 that bother the animals. In this relationship, the birds benefit and the water buffalo benefits. The birds get food, and the water buffalo gets fewer insect bites. Commensalism is when one organism benefits and the other is not affected. For example, a fly may land on a cow that is walking across a field. The fly gets a ride to a new place, but the fly 30 neither helps nor hurts the cow. The fly benefits, but the cow is not affected.

Parasitism is when one organism benefits, and the other is hurt. For example, ticks often attach themselves to the skin of other animals, such as dogs. The tick drinks the dog's blood and gets food, but the tick can make the dog sick. The tick benefits, but the dog can get hurt.

Dangers for animals

Human activities can be dangerous for animals as well as for plants. Deforestation and environmental changes such as global warming can harm them both. Many animal species are losing their habitats. Some are endangered (in danger of becoming extinct); others are already extinct. Many biologists and environmentalists think these changes in animal life are a clear warning about the future health of our planet.

Animal Communication

How do animals communicate? Do they talk? Can they have conversations and share information? Although animals do not communicate in the same way that humans do, many have their own special languages. They use their languages for different purposes. For example, animals "talk" to establish relationships with other animals: They attract mates, scare away their enemies, mark their territories, and identify themselves through communication. Many animals use several different techniques. Two common ones are auditory and tactile communication.

Auditory communication refers to the sounds that animals make. Coyotes, for example, are very noisy. They use barks, yips, and howls to mark their territory and to identify themselves, that is, to let other coyotes know where they are.

Tactile means the sense of touch, and tactile communication refers to communicating through touch. Animals use touch to show power or to form connections with others. For example, a dog may push another dog onto its back to show power and position. That is, the top dog is the "boss." Female monkeys kiss and hug their babies, and cats rub cheeks with other cats to show affection and form friendly connections.

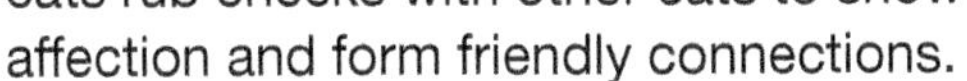

AFTER YOU READ

1 Applying what you have read

A Work with a partner. Look at the pictures and name the animals that you know. Then label each picture *V* (vertebrate animal) or *I* (invertebrate animal).

1. ____________ **2.** ____________ **3.** ____________ **4.** ____________

5. ____________ **6.** ____________ **7.** ____________ **8.** ____________

B Read the descriptions. They show different symbiotic relationships between two animals. Write *M* (mutualism), *C* (commensalism), or *P* (parasitism).

____ **1.** A shrimp digs a hole in the sand. It lives there with a goby fish. The shrimp cannot see well. The shrimp and the goby fish go outside the hole. When there is danger, the goby fish touches the shrimp to warn it. Both animals go back into the hole and both are safe.

____ **2.** The suckerfish and the shark travel together. When the shark finds something to eat, the suckerfish also eats some of the food. This does not bother the shark.

____ **3.** A tapeworm lives inside a dog. The tapeworm is safe and eats food from the dog's intestine. The dog does not get the nutrients from the food.

____ **4.** An ant protects a butterfly from enemies. The butterfly provides food for the ant.

____ **5.** A mosquito bites a bird and drinks its blood. The mosquito needs a substance in the blood to produce eggs. The bird feels uncomfortable and sometimes gets sick from the bite.

C Write *endangered* or *extinct* to describe the condition of each animal below.

__________ **1.** Dinosaurs: These animals lived on Earth until about 65 million years ago. There are no dinosaurs on our planet today.

__________ **2.** Siberian tigers: There are only about 400 Siberian tigers in the world today. Many have been killed, and many have lost their homes because of deforestation and building. These tigers may eventually disappear.

D Compare your answers to Steps A–C in a small group.

2 *That* clauses (V)

Remember that sentences in English can have more than one clause. A ***that* clause** is one type of adjective clause. It functions the same way as an adjective: It modifies, or describes, a noun or noun phrase. In the examples below, each sentence has a *that* clause and a main clause. *That* clauses always follow the nouns they modify.

main clause — adjective clause — noun phrase

A crab is a small animal **that has a shell.**

main clause — adjective clause — noun phrase

Earth is a beautiful planet **that has a lot of diverse forms of life.**

A Go back to the reading. Find sentences with *that* clauses. Underline the clauses. Then circle the nouns they modify. Compare answers with a partner.

B Complete the *that* clauses in the following sentences with your own ideas. Compare your sentences with a partner.

1. Plants are organisms that ________________.
2. Birds are animals that ________________.
3. Human activities cause environmental changes that ________________.
4. Some animals form relationships that ________________.
5. Earth is a planet that ________________.

3 Compound words (V)

Compound words are made up of two or more words. You can often guess the meaning of a compound word by combining the meanings of its parts. For example, *starfish* is made up of the two words *star* and *fish*. A starfish is an invertebrate that lives in the ocean. It is shaped like a star.

A Guess the meanings of the compound words below. Use your knowledge of the smaller words that make up these compound words.

1. backbone
2. jellyfish
3. redwood (tree)
4. racehorse
5. brainstorm
6. mountaintop
7. houseboat
8. armchair
9. toothpaste

B Use the words in the box to make as many compound words as you can.

bird	fish	rain	song	thunder
earth	flower	shell	storm	water
fall	quake	shine	sun	worm

C What other compound words do you know? Work with a partner. Make a list of compound words and share it with the class.

4 Writing about similarities and differences 

For some assignments, you will have to write about both the similarities and the differences between two or more people, places, things, or ideas. In other words, you will have to compare and contrast them. Here are some guidelines:

- Write a topic sentence that states the main idea. For example:
  X and Y are similar to, and different from, each other in several ways.
- Following the topic sentence, explain the similarities.
- Then explain the differences. First, however, write a transition sentence to signal the change in subject from similarities to differences. Here are two examples:
  X and Y have similarities, but they also have differences.
  Although X and Y have similarities, they also have important differences.
- Explain the differences.
- End with a concluding sentence that restates the main idea.

A Read the following paragraph and answer the questions below.

Sharks and jellyfish are similar to, and different from, each other in several ways. One similarity is that both animals live in the ocean. Another similarity is that both sharks and jellyfish can hurt people and other animals. A shark can bite with its sharp teeth, and a jellyfish can sting with its tentacles. Although there are similarities between these animals, there are also important differences. One difference is that sharks are vertebrates, and jellyfish are invertebrates. This means that, unlike sharks, jellyfish do not have backbones or brains. Sharks and jellyfish also have different life spans. Most jellyfish live only a few months. In contrast, most sharks live 15–20 years. These facts show that sharks and jellyfish are similar and different at the same time.

1. What two things does the writer compare and contrast?
2. What are the similarities? What are the differences?
3. Underline the transition sentence that signals the change from similarities to differences.
4. Underline all the expressions of similarity and difference.

B Compare and contrast two organisms. Choose two different animals, two different plants, or a plant and an animal. Imagine that you are going to write a paragraph and do the following.

1. Choose the two organisms you want to compare and contrast.
2. Brainstorm a list of differences and similarities between the two organisms.
3. Write a topic sentence that states the main idea.
4. Next, write a sentence that introduces the similarities and then explain the similarities.
5. Write a transition sentence. Indicate that you are changing subjects from similarities to differences.
6. Explain the differences.

5 Thinking critically about the topic Ⓐ

A Discuss the following questions with your classmates.

1. Do you know of any animals that are endangered or extinct? If so, name them.
2. Deforestation and global warming are two reasons that animals may become endangered or extinct. What are some other reasons?

B Reread the boxed text "Animal Communication." Then discuss the following questions with a partner or in a small group.

1. Have you ever seen animals communicate with each other? If so, describe the situation.
2. What are some communication techniques that animals use? Give examples.
3. What are some ways that animals communicate with people? Give examples.

C Look at the cartoon. What do you think is the main idea?

Frankly, I think we should stop trying to communicate with Humans: They're simply not smart enough to understand us

Chapter 7 Academic Vocabulary Review

The following words appear in the readings in Chapter 7. They all come from the Academic Word List, a list of words that researchers have discovered occur frequently in many different types of academic texts. For a complete list of all the Academic Word List words in this chapter and in all the readings in this book, see the Appendix on page 206.

attach	contributes	release (v)	survival
categories	establish	rigid	techniques
communicate	function (v)	structure	transport (v)

Complete the following sentences with words from the list above.

1. Trees in a small forest ________ enough oxygen in one year to keep several people alive.
2. Some language experts believe that people ________ only 30% of their messages with words, and the other 70% through nonverbal cues.
3. A newborn human baby's ________ depends on his or her caregivers.
4. It's very important to create a good business plan to ________ a new business.
5. Many people ________ an identification tag to their pet's collar. An ID tag can make it easier to find a lost pet.
6. Can you take a pineapple on an airplane? Ask security first. Many countries do not allow passengers to ________ fresh fruit across international boundaries.
7. The Queen of the Andes blooms only once in its 80- to 100-year lifetime. This fact ________ to the beauty and mystery of the plant.
8. Trainers use many different ________ to help animals learn how to behave.
9. There are five different ________ of hurricanes, from hurricanes that cause some damage to hurricanes that cause severe damage.
10. There are many types of plants, with different parts. However, the ________ of a normal house plant may include roots, stem, and flowers.

Developing Writing Skills

In this section, you will work with classification. You will also use what you learn here to complete the writing assignment at the end of this unit.

Classification

In science, many things are classified. To classify means to sort things with similar features into groups or categories. Scientists use several different classification systems. They may group by basic type or by location, composition, structure, size, weight, or origin. For example, science sorts Earth into four systems (lithosphere, hydrosphere, atmosphere, and biosphere) and living things into two basic categories (plant life and animal life). Classes are often discussed in textbooks in the following way:

Naming a class and examining its members:

- the class is named (such as "living things" or "vertebrates")
- the system used to classify the things is named (such as by type, size, weight)
- the members of the class are named and their similar features are described

It is important to note that classes include only things of true similarity.

A Answer true (*T*) or false (*F*). Use the guidelines in the box to help you.

____ **1.** Two basic types of things on Earth are living things and rocks.
____ **2.** Rocks can be classified as nonliving things.
____ **3.** Two kinds of living things are vertebrates and invertebrates.
____ **4.** Astronomers also classify planets by composition (terrestrial or gas giant).
____ **5.** Plants and human beings are in the same class.

B Look at question 5 above. How is the statement true? How is it false?

C Study these descriptions of two kinds of vertebrates. Then answer the questions below.

mammals	**amphibians**
backbone	backbone
skeleton inside body	skeleton inside body
warm-blooded	cold-blooded
do not lay eggs	lay eggs
take care of young	young take care of themselves

1. What features do mammals and amphibians have in common?
2. How are they different from each other?
3. Why are mammals and amphibians both classified as vertebrates?

D Work with a partner to complete the following activities.

1. Read the paragraph.

> One way scientists classify plants is by how they get water. The first group is vascular plants. Vascular plants have tubes in their roots, stems, and leaves. They use these tubes to carry water to all the different parts of the plant. Vascular plants stand up tall. Examples of vascular plants include ferns and trees. The second group of plants is nonvascular plants. These plants do not have roots, stems, and leaves to carry water. Instead, they soak up water from the ground, like a sponge. The water is then passed from cell to cell within the plant. Nonvascular plants grow in wet places and are not very tall. Moss and liverwort are nonvascular plants. In short, vascular and nonvascular plants have different ways of getting the water that all plants need to survive.

2. Discuss these questions:

1. What is the paragraph classifying?
2. What are the basic members of the class?
3. What classification system does the paragraph use?

3. Fill in the flowchart with information from the paragraph. Include the name of the two groups and specific examples of plants in each group.

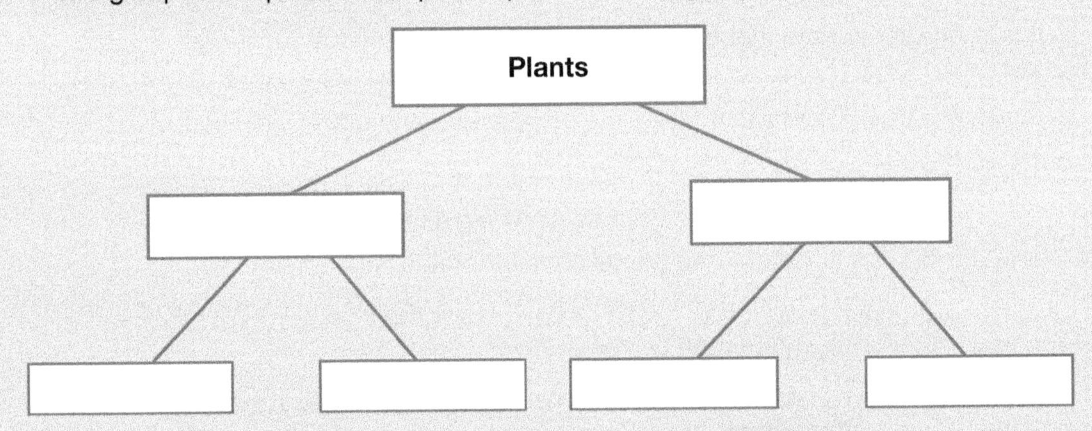

E Choose an item from the list below and present it as a class. Make a flowchart like the one above. Include examples and brief descriptions of members. Use the strategies in the box and the paragraph in Step D to help you. Look up any unfamiliar words in the dictionary. Choose a second item and repeat the process.

a. animals
b. trees
c. living things
d. nonliving things
e. computers
f. vehicles

Chapter 8
Humans

PREPARING TO READ

Thinking about the topic Ⓡ

Work with a partner to complete the following activities.

A This chapter is about human beings. Humans are unique. Why are they different from all other organisms? Brainstorm ideas with your partner. Make a list.

B The brain is one of the most important organs in the body. It controls almost everything we do. It even influences what we see. Look at the picture on the right. Describe what you see.

C Did you and your partner see the same thing in the picture? Look again. Try to find two images. If you see only one image, look at the picture from farther away.

D Discuss the images you saw in Step B with the class. Then read the information below.

The brain is always trying to understand the world around us. For example, our eyes send information to the brain, and the brain tries to understand it. Our brain tries to match the new information to information already in our memory. Usually, this is easy, but sometimes it is not. For example, look at the picture below. It looks like two different things. Your brain sees this image and then compares it to things in your memory. This can take a few minutes. Why? The brain sees a white vase against a black background. It also sees the black heads of two men, two silhouettes, with white space between them. The brain switches back and forth between the idea of a white vase and two black silhouettes. It tries to decide which one the picture shows.

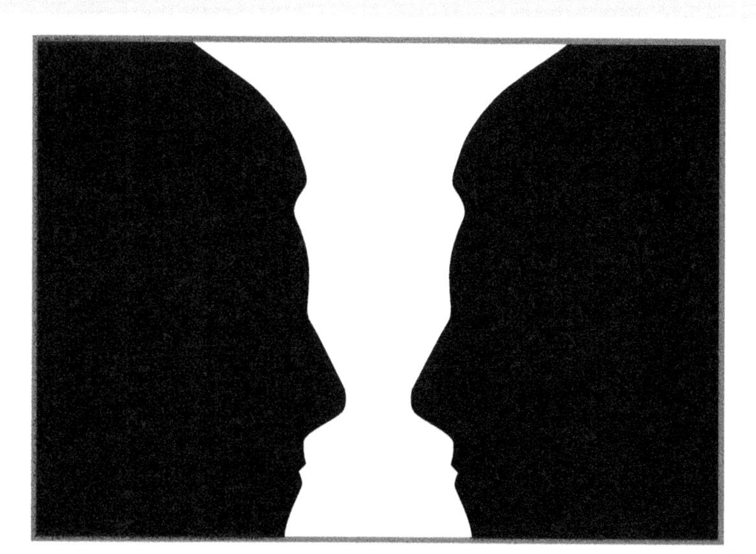

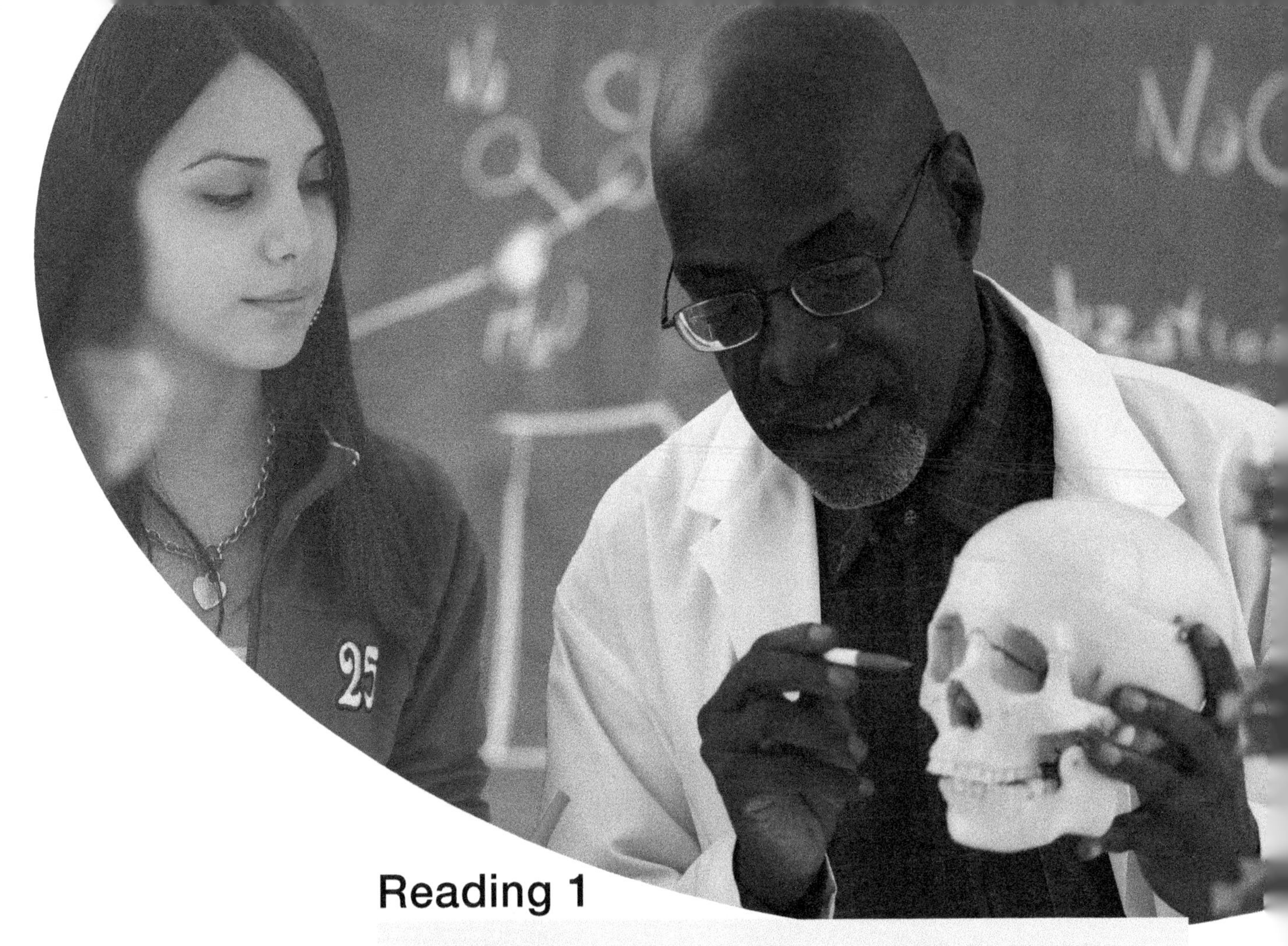

Reading 1

THE BRAIN

Just like other organisms, humans are made of cells. All cells grow, develop, and eventually die. However, as far as we know, only humans are able to think about ideas, and only humans have a sense of self (that is, a sense of uniqueness, of being an individual). An important reason for these differences is the human brain. Some animals have larger brains, but the human brain is much more complex than any other. It is the center of our thoughts, actions, feelings, dreams, and memories. Some organisms can make simple tools and communicate in basic language. However, the human brain makes it possible for us to do much more. For example, we are able to use complex language, make music, create art, and develop complex tools and technologies.

Composition, structure, and function

When scientists describe the brain, they note several key features. Surprisingly, the brain is quite small, even though it is complex. An average brain is about the size of two fists, and it weighs approximately 1.4 kilograms. Some people describe the brain as looking like a soft, pink, wrinkled rock. Others say it looks like a sponge. This important organ consists of three parts: the **cerebrum**, the **cerebellum**, and the **brain stem**.

The cerebrum is the largest part of the brain. It controls most of our thinking and speaking, and our five senses: sight, hearing, touch, taste, and smell. The cerebrum is divided into two halves, or hemispheres. The left hemisphere is important for speech, language, logic, and math skills. The right hemisphere is important for creative abilities, such as playing a musical instrument, drawing, painting, and writing books. Some people seem to use one hemisphere a little more than the other. For example, a concert pianist might use the right hemisphere more. A mathematician might use the left hemisphere more. However, in the human brain, the two hemispheres work together so that we are able to do a wide variety of things.

cerebrum the part of the brain that controls thinking, learning, language, memory, feelings, and personality

The cerebellum controls the body's movements, balance, and muscle coordination (the way the muscles work together). The brain stem is the smallest part of the brain. It connects the brain to the spinal cord and the rest of the body. It controls some of the body's basic functions, such as the heartbeat and breathing. It also handles all the messages between the brain and the body.

cerebellum the part of the brain that controls movement of the body

The processing of messages

Every second of your life, your brain receives messages from the parts of your body and the world around you. The brain interprets, or figures out, these messages, and it tells you what to think and how to act. For example, imagine that you drop a book on your foot. Your brain, not your foot, tells you what happened and if it hurts. Similarly, your eyes send messages to the brain, and the brain tells you what you see. In this way, the brain interprets almost everything you do.

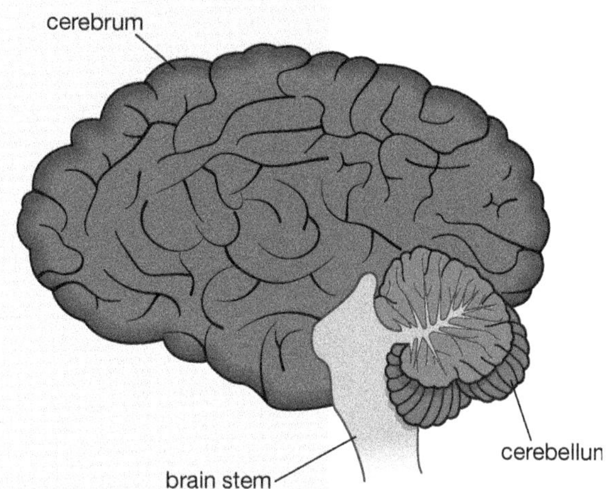

Figure 8.1 The brain

Brain research

The brain is still a mystery, but scientists are learning more about it all the time. Medical researchers are improving the treatment of brain injuries, diseases, and disabilities, such as Alzheimer's disease and autism. Others are experimenting with computer chips and paralysis. They implant computer chips in the brain, and a computer helps paralyzed people move again. Some brain researchers are applying their knowledge to the classroom. They examine how the brain learns. With this knowledge, teachers can teach more effectively, and students can learn more easily. Science has made wonderful advances in brain research, and many scientists believe that brain research is one of the most interesting and exciting areas for future study.

brain stem the part of the brain that controls basic functions, such as breathing, sleeping, body temperature, and the heartbeat

AFTER YOU READ

1 Highlighting and taking notes Ⓐ Ⓡ

Highlighting important information in a text is a helpful first step in taking notes. Try this method for highlighting:

- Read the text, but do not highlight it. Try to understand the main ideas and key points.
- Read the text a second time. As you read, highlight the main points and supporting details.

Other information:

- Do not highlight every sentence. A text with more than half of the sentences highlighted is not useful. You cannot clearly see the most important points.
- Highlight in different colors to show the types of information. For example, highlight the main points in one color and the supporting details in a different color.
- Identify the highlighted information. Note what it is and / or why it is important. Write the note in the margins of the textbook.

A Below is paragraph 6 of the reading "The Brain." Look at the student's highlighted notes. Then discuss the questions below the paragraph with a partner.

The brain is still a mystery, but scientists are learning more about it all the time. Medical researchers are improving the treatment of brain injuries, diseases, and disabilities, such as Alzheimer's disease and autism. Others are experimenting with computer chips and paralysis. They implant computer chips in the brain, and a computer helps paralyzed people move again. Some brain researchers are applying their knowledge to the classroom. They examine how the brain learns. With this knowledge, teachers can teach more effectively, and students can learn more easily. Science has made wonderful advances in brain research, and many scientists believe that brain research is one of the most interesting and exciting areas for future study.

Margin notes: brain research; brain injuries, diseases + disabilities; computer chip implants; better learning + teaching

1. Which highlighted idea is the main idea, and which ideas are details?
2. Do you agree with the student's method? Would you highlight different things? What would you highlight?
3. Are the notes in the margins clear to you? How would you note the highlighted information? Which notes would you change?

B Reread paragraphs 1–5 of "The Brain." Highlight the main ideas and supporting details. Then identify the type of information in the margins, such as "example," "key point," etc. You may find that some paragraphs have more than one main idea.

C Compare your work with your partner's.

2 Using adjectives V

Adjectives describe people, places, and things. They tell which one (this hat, that house, those shoes), what kind (color, shape, size, texture or "feel"), and how many (three books, many, some). Adjectives give details to add information and interest. For example, compare these two sentences:

- The brain looks like a rock.
- The brain looks like a **soft**, **pink**, **wrinkled** rock.

The second sentence gives a better picture of the brain. However, do not use too many adjectives. Too many adjectives make descriptions hard to read and understand.

A Read the text below. It is about a man named Phineas Gage. Underline the adjectives. Notice how they add information about the man and make the text more interesting.

Phineas Gage was a railroad employee. He was a good worker and a respected man. His employer and his employees liked him. Phineas was smart and responsible, and he was a fair boss. One day in 1848, there was a terrible accident. A heavy iron bar went completely through Gage's head. Surprisingly, he did not die, but a large section of the front part of his brain was destroyed.

Phineas's personality changed dramatically. When he went back to work the next year, everyone noticed enormous changes. Before the accident, Gage was calm, hardworking, responsible, and friendly. However, after the accident, he became angry, childish, rude, and impatient. Friends said they did not know this Phineas. For many years, scientists discussed Phineas Gage and the roles of the mind and the brain. Could damage to a certain part of the brain affect personality? Years later, scientists discovered that the front part of the brain controls personality. This explains why Gage's personality changed so much after his injury.

B On a separate piece of paper, rewrite these sentences. Add one or two appropriate adjectives to each one.

1. The brain is an organ.
2. Humans have brains.
3. There are plants on Earth.
4. Mosquitoes are insects.
5. Elephants are animals.

C Describe an organism, but do not name it. Use adjectives to make the description clear and interesting. Write three or four sentences.

D Exchange sentences from Step C with a partner. Guess what organism is being described. How did you know, or why did you not know?

3 Gerunds V

A gerund is an *-ing* form of a verb. A gerund is used as a noun. It can be the subject of a sentence or the object of a verb.

gerund as subject

Exercising keeps your body healthy.

verb gerund as object of verb

I really like **exercising.**

Be careful! Not all *-ing* forms are gerunds. Some *-ing* forms are verbs. It is important to recognize the difference, or you may not understand what you read.

-ing form as verb

We **are exercising** in the gym for one hour today.

A Read the sentences. They use *-ing* words as nouns and verbs. Underline the *-ing* words, then identify the gerunds. Write *G* (gerund) next to the sentences with gerunds.

____ **1.** Thinking is something our brains do all day long.

____ **2.** Many scientists are doing brain research.

____ **3.** Experts say that sleeping is very important for healthy brain development.

____ **4.** Hearing and seeing are two senses that the brain controls.

____ **5.** Scientists are learning more about the brain every day.

B Complete the sentences with gerunds. Use information from the reading "The Brain."

1. The cerebrum controls most of a person's ________ and ________.

2. The right hemisphere of the brain is important for creative abilities, such as ________ and ________.

3. The brain stem controls some of the body's basic functions, such as ________.

C Work with a partner. Take turns asking and answering questions with gerunds. For example, ask about places (*visit*), food (*eat*), clothes (*wear*), movies (*watch*), music (*listen*), and holidays (*celebrate*).

Which ________ do you enjoy ________-ing?

Which places do you enjoy visiting?

What is your favorite hobby?

________. (*Listening to music.*)

4 Applying what you have read

Today many scientists and educators discuss the right and left hemispheres of the brain. People use different strategies to learn. These hemispheres affect their choices. Reread paragraph 3 of the text and complete the following activity.

A Are you a right-brain or a left-brain person? Check (✓) each sentence that describes you.

____ **1.** I always look at the clock, and I like to wear a watch.

____ **2.** Before I make a difficult decision, I write down the pros (advantages) and the cons (disadvantages).

____ **3.** I often change my plans. I don't like to follow a schedule.

____ **4.** I like to draw a map to give somebody directions. I think it's easier. I can explain better with a map than with just words.

____ **5.** I learn math easily.

____ **6.** I like to draw.

____ **7.** People tell me I'm always late.

____ **8.** I learn music easily.

____ **9.** I like to have a "to-do" list.

____ **10.** I always read the directions first when I have to put something together, such as a piece of furniture or electronic equipment.

B Now look at the key at the bottom of this page. Go back to Step A. Write *L* (left hemisphere) or *R* (right hemisphere), according to the key, next to each sentence that you checked.

C Count your *L*s and *R*s. Read the information below to find the learning strategies that might work best for you.

Did you check more Ls? You may use the left half of your brain a little more than the right half. In that case, the following learning strategies might work best for you:

- Study in a quiet place.
- Work independently, not with other people.
- Memorize new words and information.

Did you check more Rs? You may use the right half of your brain a little more than the left half. In that case, the following learning strategies might work best for you:

- Work in groups.
- Draw pictures to support your notes.
- Do hands-on activities, such as conducting surveys and doing experiments.

Did you check the same number of Ls and Rs? You probably use both halves of your brain equally. In that case, a variety of learning strategies might work best for you.

Key for Step B: **1.** L; **2.** L; **3.** R; **4.** R; **5.** L; **6.** R; **7.** R; **8.** R; **9.** L; **10.** L

5 Writing a description

You will frequently need to describe people, places, things, or ideas in your academic assignments in almost every subject. For example, in science classes, you may need to describe a landform, a plant, an animal, or a part of the body.

Basic descriptions include key details, such as the height, weight, color, shape, and size of a person or thing. This means that you will use adjectives to describe nouns. A good description uses adjectives to create a clear picture for the reader.

A Read the description of a body organ from the reading. Then answer the questions below.

> The ■■■ is small and complex. An average ■■■ is about the size of two fists, and it weighs approximately 1.4 kilograms. Some people describe the ■■■ as looking like a soft, pink, wrinkled rock. Others say it looks like a sponge. . . .

1. Circle the adjectives that make the description clearer.
2. Underline the other details that help you "see" what the writer is describing.
3. What is the writer describing? Exchange ideas with a partner. Explain the reasons for your answer.

B Look at your eyes in a mirror. Then look at the eyes of some of your classmates. What do they look like? Think about their shape, color, size, and any other details that help describe them. Write some notes about the eyes on a separate piece of paper.

C Now write a paragraph that describes the human eye. Write as many specific details as you can. Include adjectives and sizes where appropriate.

D Draw a picture of an eye to go along with your description.

E Compare your paragraph with a partner's.

PREPARING TO READ

1 Thinking about the topic R

The phrases below describe bones and / or muscles. Read each description. Write *B* (bones), *M* (muscles), or *B* / *M* (bones and muscles).

_____ **1.** are hard on the outside
_____ **2.** need exercise and a good diet to stay healthy
_____ **3.** help you digest food
_____ **4.** are made of cells
_____ **5.** can get shorter
_____ **6.** (some) protect the body's organs
_____ **7.** make up a "skeleton"
_____ **8.** are used to move

2 Increasing reading speed R

A Read "The Skeletal and Muscular Systems." Use the strategies for increasing reading speed on page 68. For this task, do not read the boxed text on page 189.

1. Before you begin, enter your starting time.
2. After you finish, enter your ending time.

Starting time: ________
Finishing time: ________

Reading speed: ________

B Calculate your reading speed:

Number of words in the text (606) ÷
Number of minutes it took you to
read the text = your reading speed

Your goal should be about 80–100 words per minute.

C Check your reading comprehension. Choose the correct answers. Do not look at the text.

1. There are (*52 / 206 / 690*) bones in the human body.
2. Bones (*support the body / help the body move*).
3. Muscles (*support the body / help the body move*).
4. The human body has (*more muscles than bones / more bones than muscles*).
5. You (*can / cannot*) control all the muscles in your body.

Reading 2

THE SKELETAL AND MUSCULAR SYSTEMS

In just one day, our bodies move in thousands of ways, and we make some of those movements over and over. For example, we walk an average of 5,000 steps every day, and we usually bend our bodies 200–400 times a day. Some of our movements are unconscious, that is, 5 we do not think about them. For instance, we normally blink our eyes over 10,000 times a day. If we did not have both bones and muscles, we would not be able to walk, talk, sit, bend, blink, or smile. In fact, we would not be able to move at all.

Bones

skeleton a frame of bones that supports the body

Inside the body is a framework of 206 bones, called a **skeleton**. Bones 10 are made of living cells and tissue, and they give shape and support to the body. They are both lightweight and strong. The outside of a bone is hard and solid, and the inside has some empty spaces. The empty spaces make the bone weigh less. An average skeleton weighs only about 10 kilograms, but it is strong enough to support the body 15 and hold it upright.

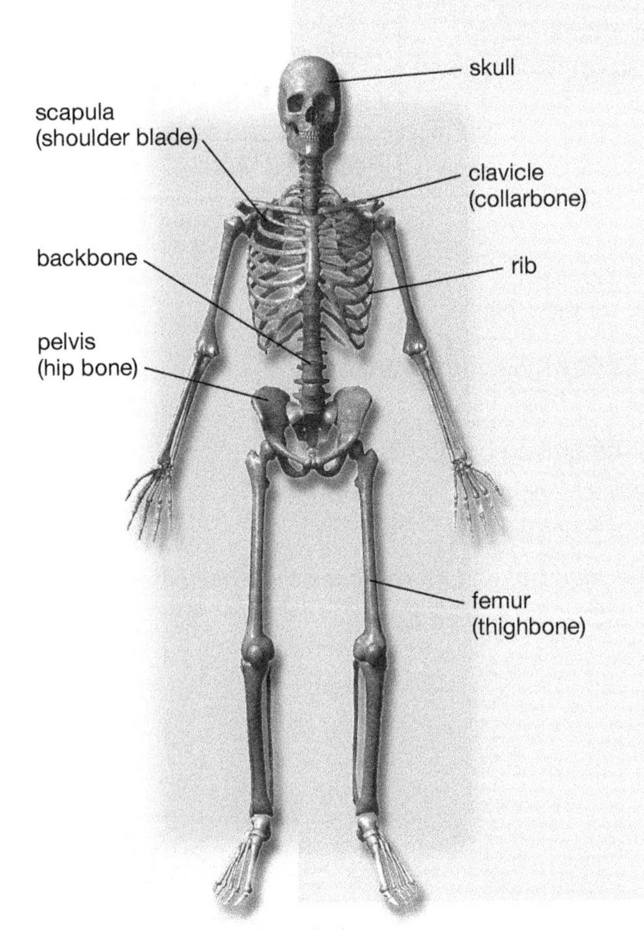

Figure 8.2 Major bones in the body

Bones have two main purposes. Some bones protect the internal organs. For example, the skull bones protect the brain, the ribs protect the heart, 20 and the backbone protects the spinal cord. Other bones, such as the femur, or thighbone, help support the body. The femur is the longest bone in the body. It is an average of 48 centimeters long, 25 and it supports the weight of the body as we walk and run.

Muscles

Muscles are also made of cells, but they are very different from bones. Muscles contract and relax to allow the body 30 to move. When a muscle contracts, it gets shorter and tighter. Some muscles contract and relax automatically. These muscles are called involuntary muscles, and we cannot control them. For 35 example, the stomach muscles contract and relax to digest food, and the cardiac muscle in the heart contracts and relaxes each time the heart beats.

Muscles that we can control are called voluntary muscles, or skeletal muscles. There are over 600 of these muscles attached to the skeleton. Voluntary muscles pull on the bones that they are attached to. In this way, they control every movement that we make. For example, the biceps and triceps muscles in our arms, and the hamstring and quadriceps muscles in our legs, are the major muscles that help us walk or pick things up. However, in order to move, we also need the smaller muscles. The smaller muscles work with the major muscles. In fact, it takes hundreds of muscles to take one step, 43 muscles to frown, and 17 muscles to smile.

Bone and muscle health

You may walk, talk, run, and smile, but you probably do not think about your bones and muscles. In fact, you probably take these parts of your body for granted unless they become injured. However, it is very important to keep bones and muscles strong. Regular exercise and a healthy diet can do this. Weight-bearing exercise like running and tennis makes your muscles stronger. This type of exercise also helps new bone tissue grow, and this makes your bones stronger, too. Eating foods that have a lot of calcium, such as milk, beans, and dark leafy green vegetables, helps your bones stay strong. Eating foods that have a lot of protein, such as lean meats and fish, helps your muscles stay strong. Take good care of your skeletal and muscular systems, and you will enjoy movement and strength for many years to come.

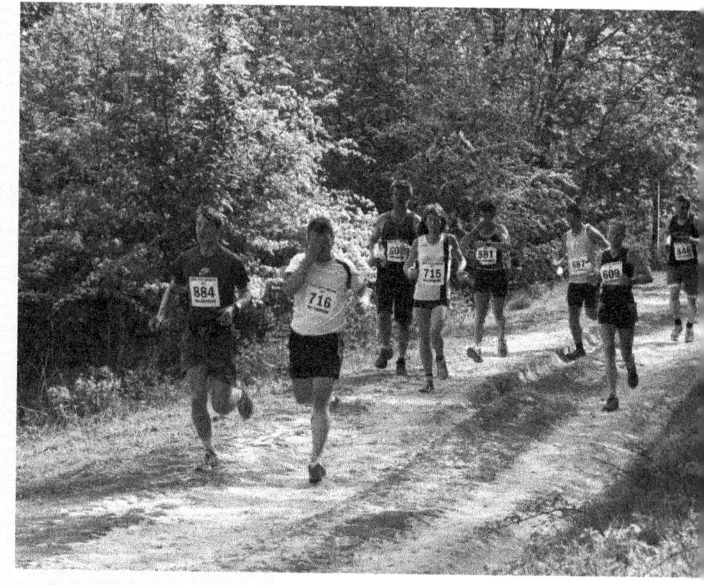

Jogging is one way to keep bones and muscles strong.

Meet Ecci

It took three years and millions of dollars, but in 2011, 25 scientists at the University of Zurich in Switzerland built a unique robot. They call the robot Eccerobot, or Ecci. Ecci is amazing. The robot has a brain that lets it learn from its mistakes. Ecci is the first robot to learn in this way. It is also the first robot to have "muscles" and "bones." The robot's muscles and bones are not real. They are made from a special kind of plastic. However, Ecci's muscles and bones allow it to move in complex ways, like a human moves. Scientists want to understand more about human movement, and they want to build better artificial arms and legs. They hope that Ecci can help teach them.

Ecci

AFTER YOU READ

1 Asking and answering questions about a text (R)

Remember that asking and answering questions about a text is a good way to make sure you have understood what you read.

Work with a partner to complete the following activities.

A Take turns asking and answering these questions about paragraphs 1–3 of the reading "The Skeletal and Muscular Systems."

1. What are some ways our bodies move every day?
2. What parts of the body allow us to move in so many ways?
3. What is a skeleton? Why is it important?
4. Describe a bone.
5. What are the two main purposes of bones?
6. Give one example of a bone and explain its purpose.

B Take turns asking and answering questions about paragraphs 4–6 of the reading.

2 Highlighting and taking notes (A) (R)

Many writers identify important information in a text in the text's margins. Taking notes in the margins about highlighted information is a useful strategy. Together, the margin notes and the highlighting focus your attention on important information in a text and help you remember it. It is often even more effective to use a separate piece of paper to write more detailed notes based on your highlighting.

A Go back to the reading. Highlight the main ideas and important supporting details in paragraph 3. Remember to use one color for the main ideas and a different color for the details. Make notes in the margins about the information you highlighted.

B Review the skills in the strategies box above and also on page 182. Go back to your highlighted information from Step A. On a separate piece of paper, write more detailed notes. Can you classify any of the information? Can you describe it in more detail?

C Compare notes with a partner.

D Choose another paragraph in the text. Highlight and take notes on the main ideas and important details.

3 Scanning for details R

Scan the reading and the boxed text to find the following information. Look only for answers to the questions. Do not reread the text.

1. How many steps do people usually take in one day?
2. How many bones does your body have?
3. How much does an average human skeleton weigh?
4. Which is the longest bone in the body? How long is it?
5. How many skeletal muscles does your body have?
6. Does it take more muscles to smile or to frown?
7. Which foods have a lot of calcium? Which foods have a lot of protein?
8. When was Ecci built?

4 Using a dictionary A R V

Read each sentence. The word in **bold** has more than one meaning. Write the correct definition and the word's part of speech in the blank. Use the dictionary and the context of the sentence to help you.

1. For example, we walk an average of 5,000 **steps** every day.
   Correct definition: ______________________________
2. Bones are made of living cells and **tissue**, and they give shape and support to the body.
   Correct definition: ______________________________
3. Some muscles **contract** and relax automatically.
   Correct definition: ______________________________
4. Some muscles contract and **relax** automatically.
   Correct definition: ______________________________
5. We also need all the smaller muscles that work with the **major** muscles.
   Correct definition: ______________________________
6. It is important to eat a healthy **diet** and exercise regularly to keep the bones and muscles strong.
   Correct definition: ______________________________

5 Words that can be used as nouns or verbs 

English has many words that can be used as either nouns or verbs. Learning both uses of these words will help increase your vocabulary.

A Read the sentences. They use the words in **bold** as either nouns or verbs:

- Bones give **shape** (n) to the body.
  Bones **shape** (v) the body.
- A **smile** (n) uses 17 muscles.
  It takes 17 muscles to **smile** (v).

B The sentences below use the words in **bold** as nouns. Rewrite each sentence and use the word in **bold** as a verb.

1. The femur gives **support** to the body when we run.

   ____

2. Doctors listen to the **beat** of a patient's heart to make sure it is healthy.

   ____

C Choose two words from the box. Look back at the text to see how the words are used. On a piece of paper, write two sentences for each word. In one sentence, use the word as a noun. In the other sentence, use the word as a verb.

control	frown	walk	work

6 Writing about the body W

The human body is a subject in many academic courses, such as medicine and anthropology. These courses look at the body in different ways. For students, writing about the body may also mean describing it in different ways. When you write about the body as a whole or as related to other living organisms, you may describe its general composition, functions, and structures. When you write about its individual systems, organs, and parts, you may need to include more details about size, weight, color, and shape in your description.

Work in a small group to complete the following activities.

A Look back at the reading "The Skeletal and Muscular Systems." Which paragraphs include descriptive writing? Underline the descriptive words and information.

B Describe a human bone. Then name and describe the system it belongs to. What is the purpose of the system? What does it do?

C Describe your hand. What does it look like? How does it feel – is it soft or hard? What adjectives would you use to describe it? Take notes on your discussion.

D Write descriptions of a body part or body system. Make a list. Do not write your name on your paper. Exchange lists with another group and guess what is being described.

PREPARING TO READ

1 Building background knowledge about the topic (R)

A Read the following information about the circulatory system.

Blood is always circulating, or moving, through the heart and around the body. It is part of the circulatory system. The circulatory system transports materials to and from all the cells. The other two parts of this system are the heart and the blood vessels. The heart pumps, or pushes, the blood through the body. When blood leaves the heart, it travels through small tubes, called blood vessels. The blood delivers oxygen, water, and nutrients to all of the body's cells. On the trip back to the heart, the blood picks up waste products, such as carbon dioxide, so that the body can get rid of them. This circulating supply of blood keeps the body working well.

B Answer these questions with a partner.

1. What does the circulatory system do?
2. Name the three main parts of the circulatory system. What does each part do?

2 Conducting an experiment (A)

In science classes, you will often conduct experiments. An experiment is a hands-on way to find out or to test information.

The blood moves through the body with each beat of the heart. You can feel this movement at pulse points, such as the neck and the wrist. Find out how fast your heart is beating. Follow the steps below.

A Hold one hand in front of you, palm up. Place the first two fingers of your other hand on the inside of your wrist. Feel for your pulse.

B When you find your pulse, count the beats for 15 seconds. Multiply this number by 4. The result is your pulse rate per minute. Record your pulse rate: ____

C When you move quickly, your body needs more oxygen-filled blood than when you are resting. This means that your heart must pump faster. Run in place for one minute and take your pulse again. Record your pulse rate now: ____

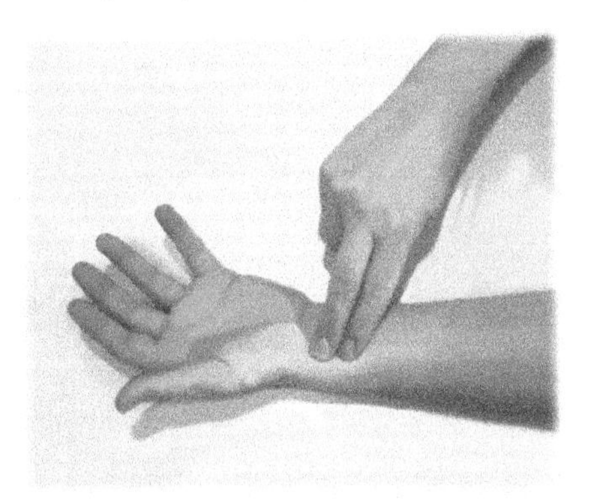

Reading 3

THE HEART AND THE CIRCULATORY SYSTEM

The heart is only about the size of a fist, but it is the most important muscle in the body. It weighs about 300 grams, and it is mostly red, just like the other muscles. The heart normally pumps, or beats, between 60 and 100 times a minute, and it delivers more than 7,500 liters of blood each day to all the cells in the body. On average, the heart beats approximately 100,000 times a day and about 2.5 billion times in a lifetime.

Features of the circulatory system

circulatory system the system that carries food and gases to the cells of the body

The heart is the center of the **circulatory system**. The circulatory system also includes blood vessels and blood. It is the body's internal transportation system. Blood travels through small tubes called blood vessels to all parts of the body. It carries the gases, water, and nutrients that people need to live and stay healthy.

Every time the heart beats, it pushes blood through the body's blood vessels. There are more than 96,000 kilometers of blood vessels inside the body. If they were stretched out, they would circle Earth more than two times. Arteries and veins are blood vessels. Arteries carry blood away from the heart to all parts of the body. Veins carry blood from the body back to the heart. Arteries and veins are connected by tiny blood vessels called capillaries.

The flow of blood through the heart

The heart is divided in half, and each half has two chambers. The four chambers hold blood that is entering and leaving the heart. The flow of blood works this way: Blood from all over the body enters the heart through the top right chamber, called the right atrium. This blood flows to the bottom right chamber, called the right ventricle. The heart then pumps the blood out of the right ventricle, through the pulmonary artery, into the lungs.

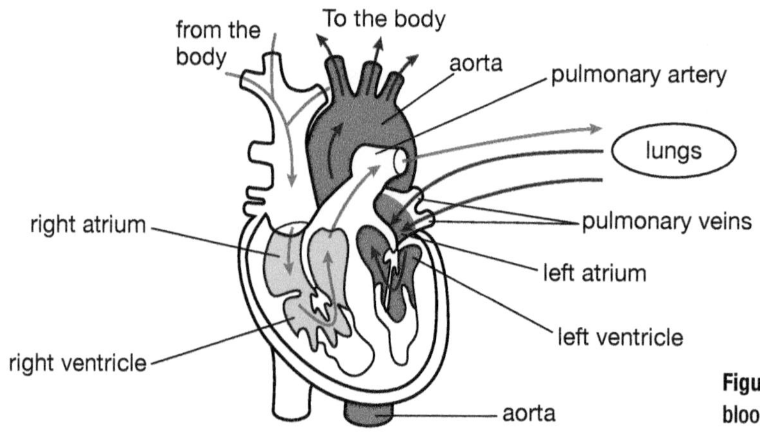

Figure 8.3 The flow of blood through the heart

The blood picks up oxygen in the lungs. Then it returns to the heart through the left atrium. Next, it flows to the left ventricle. The heart then pumps the blood out of the left ventricle into the aorta, the largest
30 artery in the body. The blood travels through the aorta and other smaller arteries to all parts of the body and delivers oxygen to all the cells. The blood then travels through capillaries to veins that lead back to the heart. From the veins, the blood goes into the right atrium of the heart to begin the process again. The whole cycle takes about 30 seconds.

Heart health

35 Your heart works hard. It started beating before you were born, and it will continue to beat for your whole life. A healthy heart has a strong heart muscle and clean, open arteries. Blocked arteries that supply blood to the heart can cause a heart attack. According to research, smoking can be dangerous to the heart. Exercise and a good diet help
40 keep the heart in good shape.

Dr. Christiaan Barnard and the First Human Heart Transplant

Louis Washkansky, a 55-year-old man in South Africa, knew he was dying. His heart was failing, and his doctors could not help him. On December 3, 1967, cardiac surgeon Dr. Christiaan Barnard performed the first human heart transplant at Groote Schuur Hospital in Cape Town. After the nine-hour operation, Mr. Washkansky had a new heart.

Dr. Barnard proved that heart transplants were possible. He became famous around the world. He continued to perform heart transplants, and he tried to solve the problems with them. The body could reject the new heart, and the medicine for this could be dangerous. Mr. Washkansky did not live long after his operation due to complications with medicine, but one patient lived for 24 years after her heart transplant in 1969.

In the 1980s, scientists developed better medicines to give patients after their heart operations. As a result, today many patients live long and healthy lives after transplant surgery. Recently, doctors have transplanted artificial hearts into some patients with very diseased hearts. An artificial heart allows a patient to survive while he or she waits for a new heart to become available. In the future, people may be able to live many years with an artificial heart.

Dr. Christiaan Barnard

AFTER YOU READ

1 Answering multiple-choice questions A R

Review the strategies for answering multiple-choice questions on pages 21 and 64. Then read the questions and circle the correct answers. Use the information from the reading. Compare answers with a partner.

1. Which sentence is not true?
   - **a.** The heart is a muscle.
   - **b.** The heart normally beats between 60 and 100 times an hour.
   - **c.** The heart pumps blood through the whole body.
   - **d.** The heart beats billions of times in a lifetime.
2. _____ is not part of the circulatory system.
   - **a.** The heart
   - **b.** A blood vessel
   - **c.** Blood
   - **d.** The femur
3. _____ is not a blood vessel.
   - **a.** An artery
   - **b.** A vein
   - **c.** A capillary
   - **d.** The heart
4. The largest artery is the _____.
   - **a.** vein
   - **b.** aorta
   - **c.** lung
   - **d.** right atrium
5. The blood is in the _____ before it goes to the lungs.
   - **a.** right ventricle
   - **b.** right atrium
   - **c.** left ventricle
   - **d.** left atrium
6. Blood gets oxygen from the _____.
   - **a.** left ventricle
   - **b.** aorta
   - **c.** right ventricle
   - **d.** lungs
7. One whole cycle of blood flow through the heart takes _____.
   - **a.** seconds
   - **b.** minutes
   - **c.** hours
   - **d.** days
8. _____ is healthy for the heart.
   - **a.** A heart attack
   - **b.** Smoking
   - **c.** Exercise
   - **d.** A bad diet
9. Which sentence is not true?
   - **a.** The circulatory system is the body's internal transportation system.
   - **b.** Each half of the heart has two chambers.
   - **c.** Your heart starts beating when you are born.
   - **d.** Blocked arteries can cause heart attacks.
10. Dr. Barnard proved that _____.
    - **a.** arteries can become blocked
    - **b.** heart transplants are possible
    - **c.** lung disease always follows heart surgery
    - **d.** people cannot live long after heart surgery

2 Sequencing R A

A Look at the steps of blood circulation. Work with a partner. Review paragraphs 4 and 5 of "The Heart and the Circulatory System" and Figure 8.3. Then number the steps 1–7.

_____ Blood flows to the left ventricle.
_____ Blood returns to the heart through the right atrium.
_____ Blood flows to the right ventricle.
_____ Blood travels through the pulmonary artery to the lungs.
__1__ Blood enters the heart through the right atrium.
_____ Blood travels through the aorta to all parts of the body.
_____ Blood picks up oxygen and returns to the heart through the left atrium.

B Complete the diagram on the right. Draw arrows to show the blood's path through the heart.

C Look back at Figure 8.3 on page 194 to check your work.

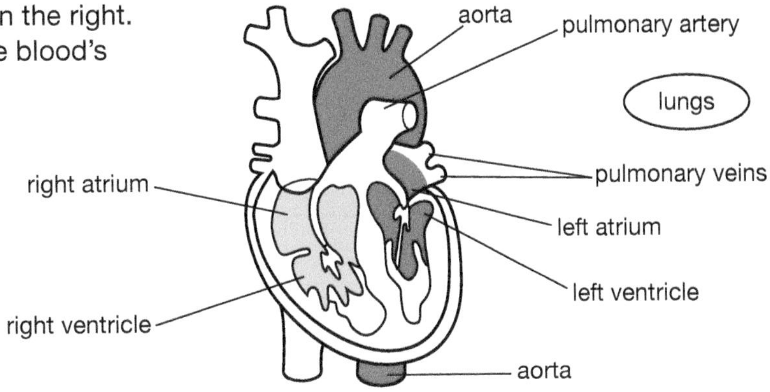

aorta
pulmonary artery
lungs
pulmonary veins
right atrium
left atrium
left ventricle
right ventricle
aorta

3 Highlighting and making an outline A R

A The reading provides information about blood vessels. Go back and highlight the main ideas and key supporting details in paragraph 3. Write short notes in the margins about this information.

B Below is an incomplete outline for part of the text on blood vessels. First, review "Making an outline" on page 164. Then complete the outline with your highlighted information in paragraph 3.

III. Blood vessels
A. Arteries: ______________________
B. ______________________
C. ______________________

C What does the text say about heart health? Highlight the important information in paragraph 6 and make notes in the margins. Organize your information on a separate sheet of paper in outline form. Start your outline this way.

VI. Heart health

A. Healthy heart = ______________________________

____.

D Compare your outlines in Steps B and C with a partner.

4 Prepositions of direction

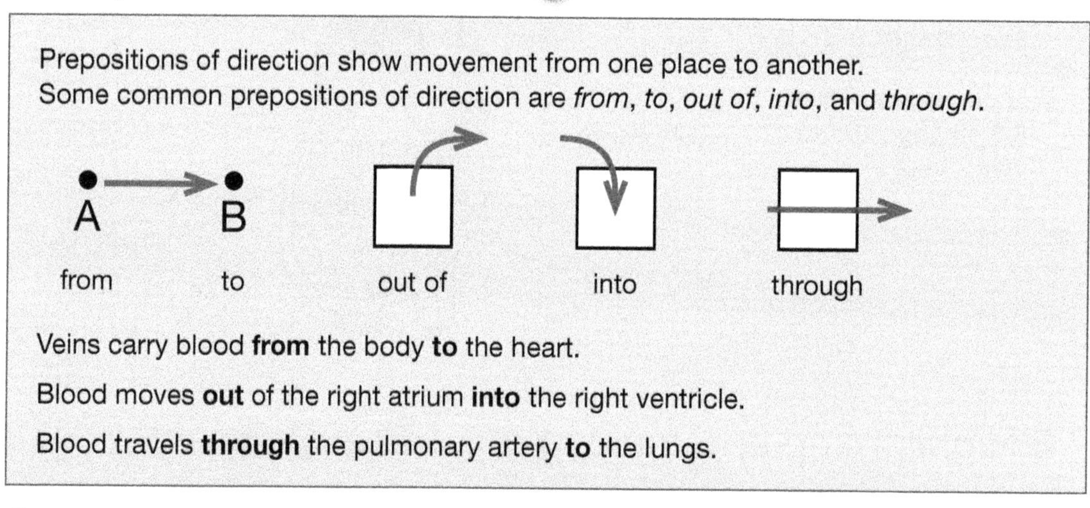

Prepositions of direction show movement from one place to another.
Some common prepositions of direction are *from*, *to*, *out of*, *into*, and *through*.

Veins carry blood **from** the body **to** the heart.

Blood moves **out** of the right atrium **into** the right ventricle.

Blood travels **through** the pulmonary artery **to** the lungs.

A Look back at paragraphs 4 and 5 of the reading. Underline the prepositions of direction. Compare your answers with a partner.

B Read the sentences. Complete each sentence with the correct preposition of direction.

1. Arteries carry blood away __________ the heart __________ all parts of the body.
2. Blood travels __________ blood vessels.
3. Blood carries oxygen __________ the lungs __________ all the cells in the body.
4. Dr. Barnard transplanted a new heart __________ the body of Louis Washkansky.
5. Blood travels __________ capillaries __________ veins.
6. The heart pumps blood __________ the left ventricle and __________ the aorta.

C Write three sentences about other parts of the human body, such as the brain, lungs, eyes, or ears. Use at least one preposition of direction in each sentence.

1. *The brain receives messages from the body through the spinal cord.*
2. ______________________________
3. ______________________________
4. ______________________________

5 Playing with words (V)

Work with a partner. Look at the words in each row. Choose the word that does not belong and explain why.

1. bones	heart	blood vessels	blood
2. gases	water	nutrients	lungs
3. right atrium	left ventricle	aorta	left atrium
4. heart	lung	brain	body
5. artery	blood	capillary	vein
6. smoking	good diet	running	swimming

6 Writing a description (W)

A Complete the chart about the key features of the heart. Use information from the reading and Figure 8.3 to help you.

Key features of the heart	Notes
size	
weight	
color	
main parts	

B Now describe the heart. Use your notes from the chart to write a paragraph about it. Remember that the topic sentence states the main idea of the paragraph and makes a point about the topic. What will your point about the heart be? Include key details and descriptive adjectives to create a good and correct picture of the heart.

Chapter 8 Academic Vocabulary Review

The following words appear in the readings in Chapter 8. They all come from the Academic Word List, a list of words that researchers have discovered occur frequently in many different types of academic texts. For a complete list of all the Academic Word List words in this chapter and in all the readings in this book, see the Appendix on page 206.

automatically	framework	interprets	reject
computer	individuals	logic	unique
coordination	internal	medical	voluntary

Complete the following sentences with words from the list above.

1. Earth is ________. It is the only planet in our solar system that has life.
2. Holding a conference for 250 people requires a lot of ________. Many departments, such as security, food services, and maintenance, must work together to make the event a success.
3. Some doors open only when someone pushes or pulls the handle. Other doors open ________; no one has to touch them.
4. Many people use the Internet to research an important ________ concern. However, it is better to ask a doctor about your health issues.
5. You do not have to attend the additional study session for Biology 100 on Friday night. The session is ________, not mandatory.
6. People in a car accident should have a doctor check them. They may not have visible marks on their bodies. However, they may have ________ injuries that you cannot see.
7. A ________ is a very useful tool. It's much faster than a typewriter, and students use it to write papers. They also use it to research information and communicate with friends.
8. Scientists often use ________ to answer their research questions. That is, they use reason, and they test their ideas with experiments.
9. The skeleton makes a ________ for the human body. This supports the body and protects it.
10. The body can ________ an artificial heart. However, now there are better medicines that help the body accept it.

Practicing Academic Writing

In Unit 4, you learned about living things. Based on everything you have read and discussed in class, you will write a paragraph about this topic.

The Human Body

You will write one academic paragraph about the human body. You will choose a part of the body, such as the heart, brain, or circulatory system. You will classify the part and then describe it in detail.

PREPARING TO WRITE

Classifying and describing

Classifying and describing are things that we often do. We classify people, animals, and things. For example, we classify people we know as friends, acquaintances, or co-workers, and then we may describe these people.

When scientists classify, they also describe. However, in science, describing is very important. It helps to identify things exactly.

A Complete the following activities with a partner.

1. Read the three paragraphs.

a.

The planet Earth is made up of four very different but interconnected systems: the lithosphere, the hydrosphere, the atmosphere, and the biosphere. The lithosphere includes Earth's crust and the top layer of the mantle. The crust is a thin layer of rock that covers the whole planet. Its thickness ranges from about 5 to 80 kilometers. The mantle is the section of Earth directly under the crust. The lithosphere is not one solid piece of rock. Instead, it is broken into many smaller pieces called plates.

b.

Invertebrates are animals, such as worms and spiders, that do not have backbones. About 95 percent of all animals are invertebrates. Many of them have a hard, protective covering, such as a shell. Invertebrates can live anywhere, but most, like the starfish and the crab, live in the ocean.

c.

Muscles that we can control are called voluntary muscles, or skeletal muscles. There are over 600 of these muscles attached to the skeleton. Voluntary muscles pull on the bones that they are attached to. In this way, they control every movement that we make. For example, the biceps and triceps muscles in our arms, and the hamstring and quadriceps muscles in our legs, are the major muscles that help us walk or pick things up. However, in order to move, we also need the smaller muscles. The smaller muscles work with the major muscles. In fact, it takes hundreds of muscles to take one step, 43 muscles to frown, and 17 muscles to smile.

2. Answer these questions for each paragraph above:

1. What class is discussed?

2. What members of the class are named?

3. How are the class and its members described?

3. Study this list of descriptive methods and classify the descriptions below.

- definition
- function
- division into parts
- numerical facts or statistics
- examples
- details of size, weight, composition, smell, sound, or other sensory data
- comparison to other things
- contrast with other things
- location within a larger group or setting

Use the list to classify the following descriptions. More than one answer may be possible.

a. There are over 600 of these muscles attached to the skeleton.

____

b. Invertebrates are animals that do not have backbones . . .

____

c. Voluntary muscles . . . control every movement that we make.

____

d. . . . but most, like the starfish and the crab, live in the ocean.

____

e. In fact, it takes hundreds of muscles to take one step, . . .

____

f. About 95 percent of all animals are invertebrates.

____

4. Go back to the paragraphs in Step A and discuss the following questions:
   1. How many descriptive methods are used? What are they?
   2. Are the descriptions exact? If so, what makes them exact?

B Now get ready to write about the human body. First, choose a topic. You might freewrite for a few minutes. Discover what topics interest you. What did you write about the most? Next, choose your topic and then brainstorm for 5–10 minutes. Include what you already know about it, questions you want to answer about it, and ideas to describe it. Make a flowchart like the one in "Developing Writing Skills" in Chapter 7 to organize your information.

C Find a picture and more information about your topic in the library or on the Internet as needed. Discuss your topic with others in small groups. Add information from your discussions to your list.

NOW WRITE

A Write the first draft of your paragraph.

B Review the paragraphs in Step A to get started. Look at your flowchart. Now start your paragraph with a clear topic sentence. Include supporting details. Use as many descriptive methods as you need to sufficiently classify and describe your topic. Use this checklist.

Are you including:

____ a topic sentence that states the main idea of the paragraph

____ major supporting details

____ minor details that illustrate the major or key support

____ a concluding sentence that restates the main idea (Be sure to make the concluding sentence a little different from the topic sentence.)

____ correct paragraph form and structure

____ vocabulary you learned in this chapter

____ descriptive adjectives and details

____ different types of descriptive methods

____ correct sentences with subjects and verbs that agree

C Give your paragraph a title.

AFTER YOU WRITE

Exchange paragraphs with a partner. Read each other's work and discuss these questions:

- Does your partner's paragraph have correct form and structure?
- What classification system did your partner use?
- Are the class and its members clearly defined and described?
- Does the paragraph include specific examples?
- Are all the ideas clear and in logical order?
- Are there any irrelevant sentences?

Revising and editing

Revising and editing is a way to improve your work. You can improve the content, form, style, grammar, and spelling of your text. Be sure to revise and edit your work to create a good paragraph.

A Revise your paragraph.

- Review your partner's suggestions.
- Think about your own ideas for revision.
- Make necessary changes.

B Edit your paragraph.

Read it again. Look for errors in spelling, verb tense or form, and other grammar mistakes. Correct any errors.

Weights and Measures

The metric system is the system of measurement that all scientists use. It is also used by people in most countries of the world. In the United States, most non-scientists use the U.S. system. Some Web sites offer a free converter that you can use to convert measurements from one system to the other.

EXAMPLES OF THE METRIC SYSTEM AND ITS EQUIVALENTS IN THE U.S. SYSTEM

The metric system is based on the number 10, and it uses different prefixes for smaller and larger units. For example, a kilometer is 1,000 meters, a centimeter is one-hundredth of a meter (.01 meter), and a millimeter is one-thousandth of a meter (.001 meter).

Units of length

Metric system		U.S. system
1 millimeter (mm)		= 0.03937 inch
10 mm	= 1 centimeter (cm)	= 0.3937 inch
100 cm	= 1 meter (m)	= 39.37 inches
1000 m	= 1 kilometer (km)	= 0.6214 mile

Units of weight

Metric system		U.S. system
1 milligram (mg)		= 0.000035 ounce
1000 mg	= 1 gram (g)	= 0.035 ounce
1000 g	= 1 kilogram (kg)	= 2.205 pounds
1000 kg	= 1 metric ton	= 2,205 pounds

Units of liquid volume

Metric system		U.S. system
1 milliliter (ml)		= 0.03 fluid ounces
1000 ml	= 1 liter (l)	= 33.81 fluid ounces
3.785 liters		= 1 gallon

EXAMPLES OF THE U.S. SYSTEM AND ITS EQUIVALENTS IN THE METRIC SYSTEM

Units of length

U.S. system		Metric system
1 inch (in)		= 2.54 centimeters
12 in	= 1 foot (ft)	= 0.3048 meters
3 ft	= 1 yard (yd)	= 0.9144 meters
1760 yd (5,280 ft)	= 1 mile (mi)	= 1.609 kilometers

Units of weight

U.S. system		Metric system
1 ounce (oz)		= 28.35 grams
16 oz	= 1 pound (lb)	= 0.4536 kilograms
2,000 lb	= 1 ton	= 907.18 kilograms

Units of liquid volume

U.S. system		Metric system
1 fluid ounce (fl oz)	= 0.007813 gallons (gal)	= 29.57 milliliters
32 fl oz	= 0.25 gal = 1 quart (qt)	= 0.9464 liters
128 fl oz	= 1 gal	= 3.785 liters

Temperature Scales

Scientists and most countries in the world use the Celsius, or centigrade, scale (°C) to measure temperature. In the United States, most people use the Fahrenheit scale (°F).

To convert temperatures from one scale to the other, use these formulas:

degrees Fahrenheit	= (°Celsius × 1.8) + 32
degrees Celsius	= (°Fahrenheit – 32) × 0.55

Appendix

Academic Word List vocabulary

access
accessible
accurate
accurately
administration
affect
affects
affected
appreciate
approach
approximately
area
areas
attach
attached
automatically
available
benefit
benefits
categories
chemical
collapse
collapsed
communicate
communicating
communication
communications
complex
computer
conduct
consists
construction
contract
contracts
contrast
contributes
coordination
create
created
creates
creative
cycle
definition
design
designs
distribution
diverse
diversity
dramatic
element
enable
energy
enormous
environment
environmental
environmentalists
equipment
establish
eventually
evidence
expand
factors
feature
features
finally
flexible
framework
function
functions
generate
generations
global
granted
guidelines
identified
identify
individuals
injured
injuries
instance
interact
internal
interprets
invisible
job
jobs
labels
layer
layers
located
location
logic
maintain
major
medical
negative
negatively
normally
occupation
occur
occurred
occurring
occurs
parallel
participate
percent
percentage
period
positive
predict
predictable
primary
process
processing
professional
range
ranges
reactions
region
regions
reject
relax
relaxes
release
releases
removed
requires
research
researchers
resource
rigid
role
series
similar
similarity
similarly
source
sources
stable
stress
structural
structure
survival
survive
surviving
symbol
tape
techniques
theory
transform
transportation
transport
unique
uniqueness
unstable
varies
vary
vehicles
voluntary
widespread

Skills Index

Adjective suffixes *93*
Answering multiple-choice questions *21, 64, 196*
Answering true/false questions *36, 157*
Antonyms *59*
Applying what you have read *130, 172, 185*
Asking and answering questions about a text *7, 190*
Asking for clarification *158*
Both...and and neither...nor *95*
Brainstorming *90*
Building background knowledge about the topic *77, 104, 116, 154, 193*
Building background vocabulary *33*
Colons, such as, and lists *114*
Comparative adjectives *10*
Compound words *173*
Concluding sentences *84*
Conducting an experiment *193*
Conducting a survey *65, 161*
Countable and uncountable nouns *67*
Cues for finding word meaning *8, 23, 166*
Defining key words *130, 165*
Describing parts *108*
Describing results *83*
Examining graphics *54, 61, 85, 110*
Examining statistics *139*
Examining test questions *107*
Gerunds *184*
Guessing vocabulary from context *107*
Highlighting *15, 88*
Highlighting and making an outline *197*
Highlighting and taking notes *182, 190*
Identifying topic sentences *60*
Identifying topic sentences and supporting sentences *72*
Illustrating main ideas *38*
Increasing reading speed *68, 140, 187*
Introducing examples *132*
Labeling diagrams *22*
Labeling a map *88*
Learning verbs with their prepositions *16*
Making a pie chart *17*
Making an outline *164*
Mapping *65*
Organizing ideas *90*
Parallel structure *94*
Parts of speech *9*
Playing with words *113, 199*
Prefixes *30*
Prepositional phrases *31*
Prepositions of direction *198*
Prepositions of location *144*
Previewing art *4, 116*
Previewing key parts of a text *12, 18, 110, 133, 161*
Previewing key terms *104*
Previewing key words *27*
Pronoun reference *44*
Reading about statistics *67*
Reading boxed texts *37*
Reading for main ideas *42, 71, 143*
Reading for main ideas and details *93*
Reading maps *31, 81*
Reviewing paragraph structure *95, 108*
Scanning *71*
Scanning for details *191*
Sequencing *58, 197*
Showing contrast *45*
Subject-verb agreement *72, 89*
Suffixes that change verbs into nouns *59*
Synonyms *143*
Taking notes *80*
Taking notes with a chart *113, 120*
That clauses *173*
Thinking about the topic *4, 18, 39, 54, 61, 77, 85, 110, 127, 140, 154, 169, 179, 187*
Thinking critically about the topic *145, 175*
Too and very *89*
Transition words *109*
Understanding averages *131*
Understanding test questions *57*
Using adjectives *183*
Using a dictionary *137, 191*
Using grammar, context, and background knowledge to guess meaning *43*
Using headings to remember main ideas *30*
Using symbols and abbreviations *120*
Using this/that/these/those to connect ideas *138*
Using a Venn diagram to organize ideas from a text *136*
When clauses *122*
Word families *157*
Words from Latin and Greek *7, 15, 121*
Words that can be used as nouns or verbs *192*
Writing a description *186, 199*
Writing about the body *192*
Writing about differences *167*
Writing about height 115
Writing about similarities 159
Writing about similarities and differences 174
Writing about superlatives 82
Writing definitions 36
Writing an observation report 123
Writing simple and compound sentences 32
Writing topic sentences and supporting sentences 73

Credits

The authors and publishers acknowledge the following sources of copyright material and are grateful for the permissions granted. While every effort has been made, it has not always been possible to identify the sources of all the material used, or to trace all copyright holders. If any omissions are brought to our notice, we will be happy to include the appropriate acknowledgements on reprinting.

Text Credits

Page 20: "Save the Rocks" is abridged and adapted from "Where Concept of a 'Pet Rock' Has Reached Its Apex" from The New York Times. August 17, 2007 © 2007 The New York Times. All rights reserved. Used by permission and protected by the Copyright Laws of the United States. The printing, copying, redistribution, or retransmission of the Material without express written permission is prohibited.

Page 35: "The Year Without a Summer" is abridged and adapted from "1816: The Year Without a Summer" published on exn.ca Discovery Channel Canada Online (now www.discoverychannel.ca) 2005 © Gloria Chang, www.gloriachang.com.

Illustration Credits

Page 13, 17, 19, 22, 28, 29, 34, 56, 58, 62, 81, 86, 87, 88, 105, 111, 120, 123, 133, 154, 165, 179, 181, 194, 197: Kamae Design

Page 5, 87, 116 (*bottom*), 153, 172, 188: Mark Duffin

Page 6: Roger Penwill

Page 38, 116 (*top*), 118, 193: Tom Croft

Page 175: © Cartoonstock.

Photography Credits

1 ©Johan Ramberg/iStockphoto; 3 (*left to right*) ©Lynette Cook/Science Source; ©Worldspec/NASA/Alamy; ©Ocean/Corbis; 5 (*right*) ©Bettmann/Corbis; 14 ©Jan Martin Will/Shutterstock; 18 (*clockwise from left to right*) ©AlbertoLoyo/iStockphoto; ©aicragarual/iStockphoto; ©Jarno Gonzalez Zarraonandia/iStockphoto; 20 ©Nikreates/Alamy; 33 ©Martin Rietze/Westend61 GmbH/Alamy; 39 (*clockwise from left to right*) ©Roger Ressmeyer/Corbis; ©Mark Downey/Masterfile; ©Nigel Spiers/iStockphoto; 40 ©Peter Essick/Aurora Photos/Alamy; 41 ©Louie Psihoyos/Eureka Premium/Corbis; 51 ©Fotosearch/Superstock; 53 (*left to right*) ©Photononstop/Superstock; ©J. A. Kraulis/Masterfile; ©Robert Harding Images/Masterfile; 55 ©Ferran Traite Soler/iStockphoto; 61 (*top to bottom*) ©Mikkel William Nielsen/iStockphoto; ©Jake Lyell/Alamy; 63 ©Lucky Business/Shutterstock; 69 (*left to right*) ©AlaskaStock/Masterfile; ©Colin Soutar/iStockphoto; 70 ©Manpreet Romana/AFP/Getty Images; 77 ©artiomp/Shutterstock; 78 ©Tom Van Sant/Geosphere/Motif/Corbis; 79 ©lorenzo puricelli/iStockphoto; 84 ©ABC Photo Archives/ABC via Getty Images/Getty Images; 91 ©Sean Davey/Documentary/Corbis; 92 ©Sipa Press/Rex Features; 95 ©Popperfoto/Getty Images; 98 (*clockwise from left to right*) ©Chawalit S./Shutterstock; ©mikehaywardcollection.com/Alamy; ©Cultura RM/Masterfile; ©redswept/Shutterstock; 101 ©Dave Reede/All Canada Photos/Alamy; 105 ©Katharina Wittfeld/Shutterstock; 106 ©Spencer Platt/Getty Images News/Getty Images; 112 ©Bill Ingalls/NASA/Handout/Corbis News/Corbis; 117 (*top to bottom*) ©Fotosearch/Superstock; ©Minden Pictures/Masterfile; 118 (*top*) ©paul prescott/iStockphoto; 119 ©Globe Trotter/Alamy; 126 ©BlueGreen Pictures/Superstock; 127 ©konradlew/iStockphoto; 128 (*left to right*) ©Lars Christensen/iStockphoto; ©Fancy Collection/Superstock; ©Rafael Ramirez Lee/Shutterstock; 129 ©epa/Corbis/Corbis Wire/Corbis; 134 (*top to bottom*) ©Chan Pak Kei/iStockphoto; ©Clint Spencer/iStockphoto; 135 ©Layne Kennedy/Nomad/Corbis; 141 ©Masterfile; 142 ©Bob Reynolds/Shutterstock; 145 ©Richard H. Cohen/Corbis News/Corbis; 149 ©William Ryerson/The Boston Globe via Getty Images/Getty Images; 151 (*top to bottom*) ©Tom Craig/Corbis Wire/Corbis; ©Gary Latham/Image Source/Corbis; ©OceanVV/Corbis; 155 (*top to bottom*) ©Leyla Ismet/Shutterstock; ©Sunny Forest/Shutterstock; ©Rich Carey/Shutterstock; 156 (*left to right*) ©Sebastian Kaulitzki/iStockphoto; ©John Durham/Science Photo Library; 160 (*top to bottom*) ©Minden Pictures/Masterfile; ©Minden Pictures/Superstock; 162 (*left to right*) ©Ursula Alter/iStockphoto; ©Photononstop/Superstock; ©Cubo Images/Superstock; 163 (*left to right*) ©age fotostock/Superstock; ©Mike Randolph/Masterfile; 168 (*left to right*) ©Robert Harding Images/Masterfile; ©age fotostock/Superstock; 169 (*left to right*) ©Nigel Pavitt/AWL Images/Getty Images; ©Cusp and Flirt/Masterfile; ©Artem Rudik/Shutterstock; 170 (*left to right*) ©Pauline S Mills/iStockphoto; ©mkurtbas/iStockphoto; 171 (*left to right*) ©Svetlana Larinav/iStockphoto; ©Tier und Naturfotografie /Superstock; 180 ©Blend Images /Superstock; 189 (*top to bottom*) ©Malcolm McHugh/Alamy; ©Attila Kisbenedek/AFP/Getty Images; 195 ©S Nicol/Historical/Corbis